Fly Ash in Cement and Concrete

by Richard Helmuth

PORTLAND CEMENT pca ASSOCIATION
Research and Development Laboratories
5420 Old Orchard Road
Skokie, Illinois 60077

Richard Helmuth is widely known for his original research papers and review articles on the physical properties and durability of portland cement concrete and the underlying structure-property relationships. He began his career as a research physicist at the Portland Cement Association R&D Laboratories in Skokie, Illinois, became manager of the Basic Research Section of the Chemical/ Physical Research Department, and since 1980 has served as Principal Research Consultant in that department at the PCA/ Construction Technology Laboratories and still actively manages major projects. He received the Sanford E. Thomson Award from ASTM Committee C-9 for his 1983 paper on ASTM standards and methods of testing fly ash and serves on several subcommittees in ASTM C-1. He has also served in the American Ceramic Society as Chairman of the Cements Division and as a member of the Board of Trustees. Dr. Helmuth holds B.S. and M.S. degrees in physics from the Illinois Institute of Technology and a Ph.D. degree in Civil Engineering-Materials from Stanford University.

Library of Congress catalog card number: 85-63562
ISBN 0-89312-085-5

SP040.01T

Printed in U.S.A.

Table of Contents

Preface

The U.S. and Canadian portland cement industry, the electric utilities, the U.S. Government, and standard-setting agencies such as ASTM are attempting to deal with the technically challenging problems posed by the need for energy conservation, environmental protection, and economy. An important aspect of these problems concerns the maximum utilization of waste materials. The amount of fly ash being generated by coal-fired electric generating stations is increasing rapidly to levels approximating the volume of portland cement being produced; however, less than 20% of the ash is used, creating a substantial waste-disposal problem. Utilization of fly ash in concrete is an effective energy-conservation measure that can improve the economy of concrete construction. To ensure effective use without creating other problems with concrete construction requires that good technical information be made available and that research needs be identified. These are the reasons for the preparation of this publication.

The technical literature concerning fly ash and its use with portland cement in concrete is very extensive; it has been reviewed many times with increasing frequency in recent years. The important annotated bibliography and review prepared in 1961 by Abdun Nur contained 275 references. Although Kokubu cited only 91 references in his 1968 review, he reported that there were almost 700 published reports on fly ash at that time. In 1980 Sersale limited his review mainly to an account of the additions to our knowledge since the review by Kokubu and Yamada in 1974; but he still cited 147 references. The PCA library card catalog lists 450 publications from 1974 to 1984, but this listing is undoubtedly far from complete. Additional recent literature was obtained by utilizing two recent computerized literature searches from the National Technical Information Service, one from Highway Research Information Service and one from the Smithsonian Science Information Service, as well as previous reviews and the PCA library resources.

Many of these reports do not deal directly with the use of fly ash in concrete, although they do contain valuable information obtained in studies concerned primarily with combustion or pollution control. This review is necessarily selective because of the vast size of potentially applicable literature. Nevertheless, this writer believes that enough of the literature has been examined to ensure that items of importance have not escaped attention.

Before proceeding to examine the state of our knowledge of fly ash characteristics and a number of technical problems in the use of fly ash in concrete that are apparent from the published literature, it is well to note that our knowledge is still incomplete. The nature of report writing is to point out what has been done; less emphasis is placed on what needs to be done or on stating clearly the problems that remain unsolved. This particular approach is adopted even by many of the review articles; some are written in an extremely positive way, pointing out the advantages of the use of fly ash in concrete with apparent enthusiasm. Similarly, successful projects are often described in print, whereas many small failures may go unrecorded. Despite these limitations, it is hoped that this book is sufficiently complete and objective to be useful to a broad spectrum of readers.

Acknowledgments

The author gratefully acknowledges assistance from the many individuals and organizations who have generously provided information and guidance.

The book was written as the initiation of a new basic research project proposed by Eivind Hognestad at the Portland Cement Association R&D Laboratories after extended discussion with Birger Warris concerning the accelerating trend toward portland-pozzolan cements and fly ash additions.

Of the many people who provided help with the work, special thanks are due to Bryant Mather and Katharine Mather at the U.S. Army Engineers Waterways Experiment Station in Vicksburg, Miss.; Geoffrey Frohnsdorff and Paul Brown at the U.S. National Bureau of Standards, who helped get the work underway; Fred Cohrs, Hank Fischer, Nate Greening, Harry Harris, Peter Hawkins, John Hickel, Jr., Jean Laneuville, and Jan Skalny, each of whom reviewed one or more chapters and made valuable suggestions for improvements. From PCA staff, Cynthia Spigelman provided resourceful assistance in the Library, and Paul Klieger, Stewart Tresouthick, Ellis Gartner, Donald Campbell, David Stark, and Muhammad Bhatty each reviewed portions of the manuscript. Eileen Avner produced beautifully typed manuscripts at each revision. Nancy Ady's contribution to the work would be difficult to overestimate; she has checked, edited, and corrected every draft with care and great skill. Finally, my wife Karen's forebearance and support made the work go easily, both in the office and at home.

Abstract

Fly ash from pulverized coal combustion poses a major solid waste disposal problem in the United States and Canada; however, fly ash could become much more widely utilized in blended cements or with portland cement in concrete if technical characterizations were improved to ensure good performance of the concretes. Some fly ashes can even be used to improve concrete properties.

Fly ashes are fine powders formed from the mineral matter in coal. They consist mainly of glassy spherical particles as well as residues of hematite and magnetite, char, and some crystalline phases formed during cooling. Properties of fly ashes vary significantly with coal composition and plant operating conditions. Some fly ashes reduce the water requirement for flow when used as partial replacements for portland cement in paste, mortar, and concrete. The conventional explanation for this phenomenon in terms of spherical particle shapes is probably incorrect; the deflocculation effects appear to be more important. ASTM mortar flow tests may yield misleading results for some fly ashes.

Hydration of cement is usually accelerated by fly ash after some initial retardation, and the hydration chemistry is somewhat different in mixtures containing fly ash. The silica-alumina glassy particles react with the calcium released during cement hydration to form additional calcium silicate hydrate and to reduce the CaO to SiO_2 ratio of the calcium silicate hydrate gel. Although ettringite and calcium monsulfoaluminate form in the early reactions, the CaO to Al_2O_3 ratios of the calcium aluminate hydrates are also reduced when fly ash is present so that hydrogarnet and gehlenite hydrate are formed.

The most important factors determining the reactivity of fly ashes are the fineness, glass content, and silica-plus-alumina contents. The common methods used to determine the fineness of fly ashes need to be improved. The ASTM C311 standard test for pozzolanic activity (strength) with lime is specified to be performed with very lime-deficient proportions for highly reactive ashes; the test for pozzolanic activity with portland cement is specified to be performed with proportions that should permit complete reaction of highly reactive ashes and yield stable hydrates. The results of such accelerated tests do not, however, correlate well with the results of tests with long-term curing. Some rapid chemical methods, such as the rate of dissolution in 2% HF, appear to correlate well with cementing efficiency factors. The foam index test for air-entraining agent requirement appears to be reliable, but most tests for soundness and durability need to be better understood and improved. ASTM Class F (pozzolanic) fly ashes usually improve sulfate resistance and reduce alkali-aggregate reaction expansions. Some ASTM Class C (cementitious and pozzolanic) fly ashes may reduce durability.

Further research is required to improve test methods that characterize fly ashes and to provide a basis for revision of specifications to assure satisfactory performance.

Notation

The following notation, widely used by cement chemists, is used throughout the text:

A = Al_2O_3	F = Fe_2O_3	N = Na_2O
C = CaO	H = H_2O	S = SiO_2
$\overline{C}$ = CO_3	K = K_2O	$\overline{S}$ = SO_3
	M = MgO	

Compounds of interest are then more simply represented, for example:

tricalcium silicate = $3CaO \cdot SiO_2$ = C_3S
calcium hydroxide = $Ca(OH)_2$ = CH
calcium silicate hydrate gel (indefinite composition) = C-S-H

The lime-to-silica ratios are expressed in several ways:

C/S as a molar ratio,
CaO/SiO_2 as either a weight or mole ratio,
or Ca/Si as an atom ratio

Chapter 1
Introduction

Objectives

This review of the literature on fly ash was prepared to identify problems and research opportunities concerning the characterization of fly ash for its use with portland cement in concrete. Despite the fact that fly ash has been used in concrete for more than 50 years and extensive literature on the subject exists, it is apparent that such use has been based largely on empirical relationships of limited validity. There have been persistent problems, therefore, concerning the variability of fly ash and the difficulty of characterizing it for predictable performance in concrete. These problems are further complicated by the fact that much of what has been learned concerning bituminous coal fly ashes does not apply to subbituminous coal and lignite fly ashes, sources of which are being rapidly developed in the Western United States.

Despite the extensive literature on the properties of cement and concrete made with fly ash, the essential chemistry and physics of these systems are not well defined. Hence, the development of test methods and specifications has had an inadequate scientific basis and cannot provide a sound basis for predicting performance. Some tests, in fact, are no more than accelerated performance tests of trial mixtures; these can be very time-consuming and still not be reliable. It is evident that well-conceived programs of research are necessary. To develop effective programs, a thorough review of previous research was deemed essential.

The focal points of this work are (1) the physical and chemical characteristics of fly ashes and the methods used to determine them and (2) the interactions between portland cements and fly ashes in fresh and hardening mixtures and how these interactions are affected by the characteristics of the fly ashes.

The objectives of this study are (1) to provide the reader with a good summary of the technical literature on fly ash, (2) to show both the adequacy and the limitations of current test methods and specifications with respect to the use of fly ash with cement in concrete, and (3) to identify research needs. The results should then provide a sound basis for the improvement of existing test methods or the development of new methods and specifications to assure predictable and improved performance of fly ash with portland cement in concrete.

Historical Background

Both the ancient Greeks and Romans were aware that certain volcanic deposits, when finely ground and mixed with lime, sand, and water, yielded mortars of superior strength which also resisted deterioration in fresh or salt water.[(7B)]*

---

*Superscript numbers in parentheses (with or without a lowercase *a* or *b*) designate references listed at the back of this publication.
Superscript numbers followed by a capital B indicate listings in the bibliography.

Roman builders used the red or purple volcanic tuff found near the Bay of Naples. The best variety of this earth was obtained near Pozzoli and acquired the name of Pozzolana. This designation has since been extended to the whole class of minerals with similar properties. In the United States the term pozzolan is used to describe natural materials as well as some calcined clays and shales.[1] The fine fly ash produced from the burning of powdered coal has been described as an artificial pozzolan, along with water-quenched boiler slag and precipitated silica.[2]

The earliest report on the use of fly ash as a pozzolan in concrete appears to date back to 1914, but the first comprehensive study was that described in 1937 by R. E. Davis and his coworkers at the University of California.[3,4] Research from that time was paralleled by research on economical fly ash collection and handling systems. These studies stimulated continuing research and utilization of fly ash with concrete. The decision by the U.S. Bureau of Reclamation in 1948 to use large amounts of fly ash in the construction of the Hungry Horse Dam had a major impact. Extensive testing and research was conducted before the decision was made, and the work was well documented in government reports and other publications.[36B] This was the fourth largest dam in the world at the time and was estimated to require 120,000 metric tons of fly ash and 155,000 metric tons of portland cement (i.e., a mixture that was 44% fly ash). The concretes actually used had higher portland cement contents, the amounts of fly ash being 24% of the cement plus fly ash for the exterior concrete and 32% for the interior. After publication of the performance data for this major project, the use of fly ash in concrete for hydraulic and various other structures spread over the United States and around the world.

ASTM tentative specifications for fly ash (ASTM C350-54T) and portland-pozzolan cement (ASTM C340-54T) were adopted in 1954. Standards for fly ash were adopted in Japan in 1958, in the U.S.S.R. in 1963, and in Great Britain in 1965. France adopted a standard for fly ash cement in 1959, as did Japan in 1960.[3]

Fly Ash Utilization

Each year since 1966, accurate data on the amounts of ash produced and utilized in the United States have been collected by the Edison Electric Institute and, since 1973, by the American Coal Ash Association (formerly the National Ash Association). Results of these surveys along with some earlier data showed that the amount of fly ash collected annually in the United States increased steadily from 6 million metric tons in 1950 to 37 million metric tons in 1974.[5] For comparison, U.S. portland cement annual production increased from 38 to 61 million metric tons from 1950 to 1975;[6] in 1984 the annual cement clinker production capacity was about 80 million metric tons. Results of the fly ash collection surveys up to 1984 are shown in Fig. 1.1.[7a,7b] Fly ash collection was expected to increase rapidly during the 1980's to about 70 million metric tons annually by 1985,[8] due mainly to construction of additional coal-burning plants,[9] higher collection efficiency, and the higher ash contents of lignite and western coals; but this expectation has not yet been realized.

More than 300 million metric tons of fly ash have been accumulated as solid wastes.[10a] Only about 3 million metric tons, or 6% of that collected, were used by the cement and concrete industries in 1979 but this increased sharply in 1984 to 5 million metric tons.[7a]

An estimated 70% to 90% of this amount was used by ready mix concrete producers.[10b] Less than 1% of the fly ash collected in 1979 was used in the production of ASTM Type IP Blended Cement.[21B] Total fly ash utilization, including other uses such as soil stabilization and asphalt filling, increased from 8% in 1966 to 13% in 1976 to 20% in 1984.[7b] By contrast, 24% of the fly ash collected in 1978 in France was utilized in cement and concrete.[11] The reasons for the relatively low level of utilization of fly ashes in concrete in the United States have been discussed in terms of fly ash variability, the restrictiveness of some specifications (e.g., those based on tests for sulfate resistance), the inadequacy of test methods, and the need to develop additional durability and engineering performance data.[10a,12,50B] Many other countries have similar concerns.[11,13,50B]

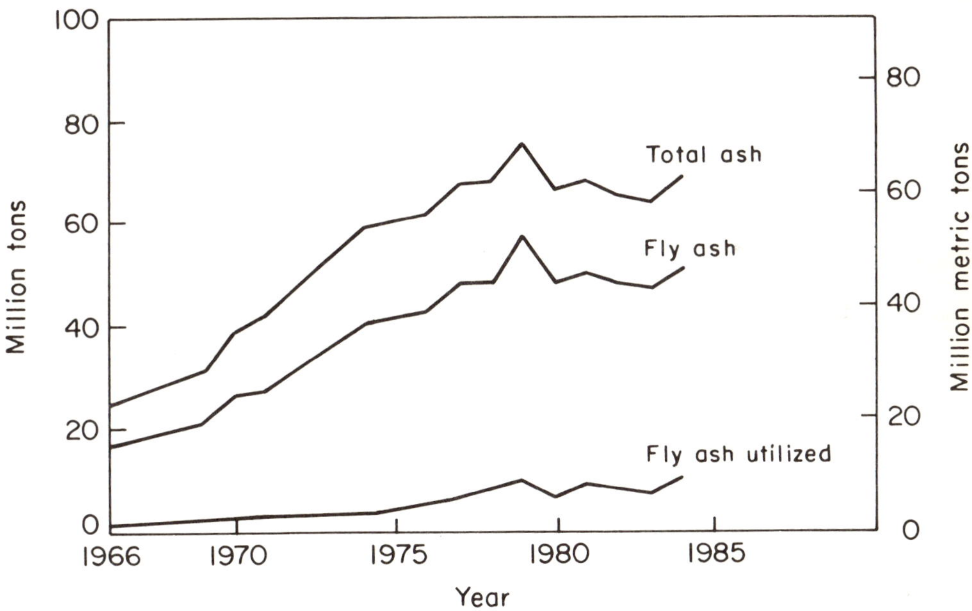

1 ton (2000 lb) = 0.9072 metric tons

Fig. 1.1 Ash collected by U.S. electric utilities and fly ash utilized. Source: American Coal Ash Association[7a,7b]

Economic and Regulatory Background

The recent trends just discussed are related to the cost of fuel and the effect of environmental protection measures. Long before the establishment of the U.S. Environmental Protection Agency (EPA) and federal air quality standards, ash collection systems were being installed in coal-fired electric generating plants. However, the establishment of limits on particulate and SO_2 emissions has increased the volume of fly ash and scrubber sludges collected. Disposal of these materials is not only expensive but also creates other kinds of pollution problems.[35B] Drainage of effluents from ash disposal ponds threatens fresh water supplies because trace quantities of toxic elements are present in the fly ash.[14] Hence, the electric utility industry and government agencies are actively seeking means to increase the use of these materials in ways that do not pose a threat to our environment.

The problem is compounded when costs of petroleum and natural gas increase and cause the electric utilities to increase their reliance on coal. The problem is also aggravated by the slowdown in nuclear power development.

To some extent these problems may be relieved by improvements in combustion equipment such as cyclone-fired boilers, fluidized bed reactors, and "sticky wall" collectors.[15,16] These would greatly reduce the amounts of fly ash produced but would increase the volume of bottom ash and slag (see "Combustion and Ash Formation" in Chapter 2). At present these alternatives must appear attractive to the utility industry, and research is being done to improve these techniques. However, it has already been demonstrated that much more fly ash could be employed as a cementing ingredient in portland cement concrete. As produced, most fly ashes are of cement fineness; high-quality fly ashes require little or no processing. Because portland cement production is a very energy-intensive process, the use of fly ash substantially reduces energy requirements as well as the cost of concrete construction.

The energy economics of the cement industry have been analyzed by Spellman.[17a] He points out that the need for energy-efficiency will also boost the demand for the most energy efficient construction materials. Although cement manufacturing is very energy intensive, concrete as a building material is far more energy efficient than most other construction materials. Thus, an increasing cost of energy would price more energy-intensive materials out of markets that can be satisfied by concrete, increasing the demand for cement and concrete. He foresaw severe cement shortages in the 1980's unless new production facilities were built or supplementary materials such as fly ash were used. Much production capacity existed in older plants that were not energy efficient. The average fuel consumption of U.S. cement plants in 1978 was 7.1 GJ/t (6.1 million Btu/ton) but ranged from 4.1 to 13.7 GJ/t (3.5 to 11.8 million Btu/ton). Energy efficiency improvements reduced average fuel consumption to 5.3 GJ/t (4.5 million Btu/ton) in 1985.[17b]

The cement industry has almost completely converted to coal as its primary fuel and is using new energy-efficient kiln systems to conserve fuel. About 40% of production capacity is still wet process which, on the average, is considerably less energy efficient than dry process and also more expensive to convert to high-efficiency kiln systems.

For these reasons and others associated with pollution controls and foreign competition, a number of plant shutdowns have occurred and more are anticipated. Spellman estimated that capital investments in the billions of dollars will be needed to upgrade or replace existing facilities and provide the capacity required to prevent cement shortages in the next decade. For the relatively small domestic cement industry, this is a difficult task.

In addition to changes in sources of capital and in public policy regarding tax credits and regulatory matters, Spellman called attention to two other possibilities that could have a major impact on energy usage within a relatively short time and with low capital requirements, viz., (1) the judicial use of pozzolans and granulated blast furnace slags as cementitious components, and (2) the use of waste fuels and refuse as supplementary kiln fuels. The use of fly ash in the production of blended cements not only reduces energy consumption per ton of cement but also has the effect of increasing plant capacity.

It appears that the cement producers and the electric utilities are being brought together by mutual economic and environmental concerns.

For several years, various agencies of the U.S. Government have sponsored research on energy conservation and environmental matters related to cement manufacture and use.[18–21] A very recent development has been the direct intervention of the EPA concerning the use of fly ash in the cement and construction industries. The growing realization that fly ash is becoming a monumental solid waste disposal and environmental problem prompted the EPA to develop and issue its proposed "Guideline for Federal Procurement of Cement and Concrete Containing Fly Ash."[8] Under Section 6002(e) of the Resource Conservation and Recovery Act (RCRA) of 1976, the EPA is instructed to prepare guidelines to assist procuring agencies in complying with the law. The EPA's regulatory strategy was explained in a lengthy preamble to the proposed guideline.

Section 6002 of RCRA requires procuring agencies using appropriated federal funds to purchase items composed of the highest percentage of recovered materials practicable. The EPA targeted a number of recovered waste materials, including fly ash, and issued a proposed guideline in November 1980. The rationale for choosing to issue the guideline was based upon four considerations: (1) fly ash is a significant solid waste problem, (2) feasible methods of recovery exist, (3) fly ash has technically proven uses in cement and concrete, and (4) federal purchasing power for concrete, both direct and indirect, is substantial.

Of particular importance is the section of the preamble on Specifications, which states:

> Section 6002 of the Act requires revision of all specifications for cement and concrete for which Federal agencies have drafting and review responsibility so that they no longer exclude the use of fly ash—and so that they incorporate the use of fly ash to the maximum extent.

In addition, the section on Purchasing states:

> EPA is designating fly ash used in cement and concrete as a product area for which procuring agencies must exercise affirmative procurement actions under Section 6002 of RCRA."

However, the EPA also recognized that satisfactory performance, availability of suitable fly ash, satisfactory competition, quality control, and lack of knowledge of the proper use of fly ash by many contractors were important factors that prevented the EPA from *requiring* the use of fly ash in all cases. Instead, the rules were written to ensure that specifications would *allow* the use of fly ash where technically feasible, and the agency recommended that alternative bidding procedures be used in federal purchasing.

Since almost one-half of total U.S. cement consumption is in public construction projects (public buildings, public works, transportation, etc.) and since federal funds account for nearly two-thirds of the funding for public construction nationally, approximately 23 million tons of cement are purchased annually, either directly or indirectly with federal funds. To monitor implementation, certification of fly ash content will be required in contracts.

With respect to quality control and performance, the guidelines also recognized that:

> ASTM specification C618 may not be sufficiently stringent to ensure all ashes which comply are really suitable for general use as mineral admixtures in concrete, particularly when chemical admixtures such as air entraining agents are to be used.

Two major problems concerning specifications for fly ashes are (1) the variability of fly ashes, even from a single source, and (2) the time required to demonstrate adequate performance of materials that are commonly expected to provide dependable service for 20 to 50 years or more.

At a public hearing on January 8, 1981, representatives of the cement and construction industries made presentations pointing out technical and economic problems that may result from the guideline.[22] On the basis of formal testimony and on 161 written comments submitted in response to the EPA's request for comments, the guideline was revised. The final guideline was issued with an effective date of February 28, 1983.[8] The preamble for the final guideline addresses specific points raised by commentators and clarifies the revisions made. The principal revisions are summarized below:

1. The EPA's distinction between allowed and required use of fly ash was clarified, and the provision to "require the use of fly ash to the maximum extent practical" was deleted.
2. Proposed longer curing periods in performance standards (e.g., 90 days instead of 28 days for airfield pavement strengths) were eliminated.
3. Minimum cement content specifications for concrete were changed to consider portland cement *plus* fly ash as the total cementitious component.

4. Quality control guidelines were strengthened by the following provisions:
   a. Procuring agencies should expect suppliers of blended cement, fly ash, and concrete to demonstrate (through reasonable testing programs or previous experience) the performance and reliability of their product and the adequacy of their quality control programs. However, procuring agencies should not subject cement and concrete containing fly ash to any unreasonable testing requirements.
   b. In accordance with standard industry practice, fly ash suppliers should be required to provide to users a statement of the key characteristics of the fly ash supplied. These characteristics include its chemical constituents, loss on ignition (LOI), and fineness of the matter. These characteristics may be stated in appropriate ranges. Other characteristics should be requested as needed by the procuring agency.

In addition, the guideline states that maximum use should be made of existing federal specifications and voluntary consensus (ASTM) standards.(21B,22B)

Among the cement and construction industries and the electric utilities are a growing number of brokers in fly ash. The EPA regards the role of these brokers, that of selecting fly ash for quality and uniformity, as important and necessary for wider acceptance and use of these byproducts.(23) Some of these brokers provide technical information to users (principally ready mix concrete producers) and technical service when problems arise.(24) The ready mix concrete producers play a crucial role in the selection and proper use of fly ash, in some cases arranging for testing at the power plant to preclude shipment of unsatisfactory ash.(25)

In some areas fly ash is widely used with relatively few problems. In Chicago the use of fly ash in concrete in large commercial buildings has been notable, including that in Water Tower Place, the world's tallest (262 m, 76 stories) reinforced concrete building, and in Sears Tower, the world's tallest (442 m) building.(26) In other areas, problems may develop, as with the drastically reduced sulfate resistance of concrete made with some fly ashes in the Western United States.(27) An important issue confronting the cement and construction industries is how to prevent such problems involving the use of fly ash and thereby assure customer satisfaction with concrete construction. One way is to produce fly ash-blended cements with strict quality control at the cement plant.

Kriner(28) has attempted to show by a simple cost calculation that cement manufacturers cannot produce at current prices Type IP blended cements(21B) with fly ash that would be cost competitive with concrete products using the same fly ash as an admixture. However, as Berry and Malhotra(29) have pointed out, many authors who discuss the economic benefits of fly ash as an admixture have failed to consider the cost of handling and proportioning an extra component of the concrete. Such economic considerations must depend on such matters as shipping costs and the size of the operations. In very large public works batching plants, the use of fly ash as an admixture has been demonstrated to contribute both to economy and quality. On large dam projects in Japan, fly ash has been blended in a wet slurry, a process which has both enhanced the uniformity of the fly ash and improved the economy of its production.[3] For smaller projects, blended cements from cement plants should permit more consistent quality control than is likely to be attained in smaller, scattered ready mix concrete plants and construction sites. In either case, quality control depends upon the test methods that are used.

Technical Background

The following chapters concentrate on the chemistry and physics of the simpler constituent systems present in portland cement concretes, with and without fly ash. Much information obtained in tests of concretes has been deliberately omitted because of the difficulty of identifying causes and quantifying effects in such complex systems. However, in order to provide some background for understanding the relationship between the characteristics of fly ashes and the properties of concretes, the remainder of this chapter will be devoted to some of the major effects of using fly ashes in concrete and current standard specifications for such use.

A recent review by Berry and Malhotra[29] provides an excellent summary of the major effects of the use of fly ash in concrete. The inclusion of fly ash in concrete mixes affects all properties of concrete. It has a major influence on the rheological properties of fresh concrete, strength development, finishability, permeability to water and aggressive solutions, and durability. Some fly ashes reduce the water requirement for a given consistency of the concrete. Fly ashes react as pozzolans to form cementitious hydrates similar to those formed by portland cement, although much more slowly. To achieve satisfactory early strength, the concrete mixes must be proportioned accordingly, taking advantage of the reduced water requirement with fly ash. Fly ashes are used to achieve very high ultimate strengths and very low permeability after prolonged curing. The use of fly ash in practicable proportions does not significantly affect the drying shrinkage of concrete. Resistance to chemical attack, such as that caused by sulfates, can be improved by the use of some fly ashes, and expansion and cracking due to alkali-aggregate reaction can often be very well controlled. Resistance to damage caused by freezing and thawing may also be lowered by the use of fly ash.

Proportioning of Concrete with Fly Ash

Three basic mix proportioning approaches have been developed primarily for use with pozzolanic fly ashes:

1. Replacement of a portion of the cement with fly ash often results in lower early strengths. For replacements up to about 20% fly ash, the strengths at ages beyond about 3 to 6 months will often exceed the strength of the portland cement concrete, as indicated in Fig. 1.2.[30]
2. The addition of fly ash as fine aggregate without reduction of the portland cement content generally increases strengths at all ages, and in some cases improvements may equal or exceed those achieved with an equal addition of cement.[31]

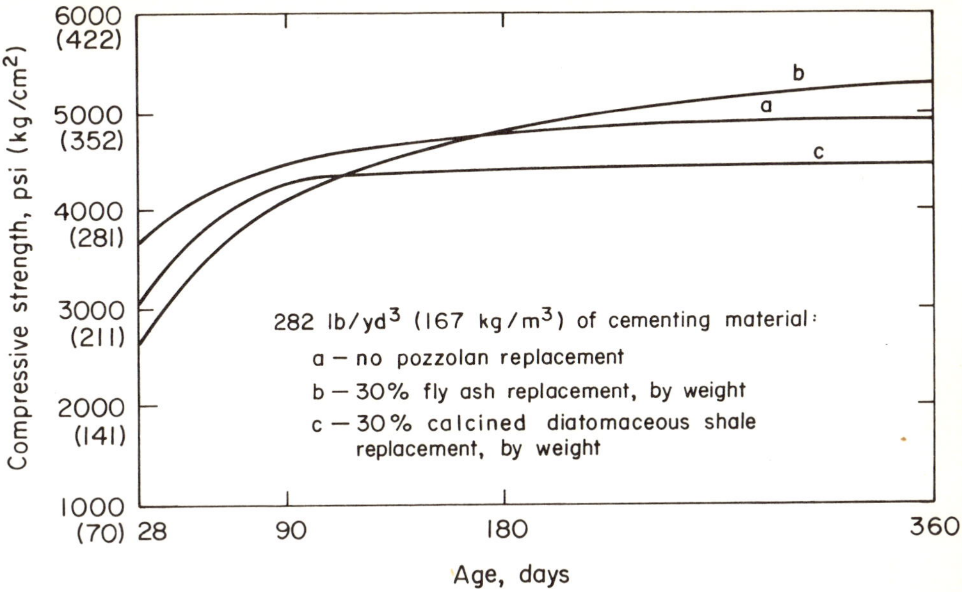

1000 psi = 6.894 MPa = 70.3 kg/cm²

Fig. 1.2 Comparison of rates of strength development commonly reported for plain portland cement concrete and concrete containing pozzolanic (Class F) fly ash as a partial replacement for portland cement. (Ref. 30)

3. Partial replacement of cement, water, and fine aggregate, so that the total cement-plus-fly ash content is increased, leads to the concept of optimization of proportions for both strength development and economy.[32–36]

Smith[34] has also introduced the concept of cementing efficiency, *K*, of an ash as a means of characterizing fly ashes for their use in concrete. In its effect on the water-cement ratio required for a given strength, a given weight *f* of ash is equivalent to a weight *Kf* of cement.

Thus, for a concrete containing weights *w* of water, *c* of cement, and *f* of fly ash, the effective water-cement ratio for equal strength is

$$(\mathrm{w/c})_e = \frac{\mathrm{w}}{\mathrm{c} + \mathrm{Kf}} = \frac{\mathrm{w}}{\mathrm{c}}\left[\frac{1}{1 + \mathrm{K}\frac{\mathrm{f}}{\mathrm{c}}}\right] \tag{1.1}$$

where *K* has a value of about 0.25 for the ashes used.

Instead of the cementing efficiency, *K*, Dodson used the ratio of cement factor, C.F. (kg/m^3), to the water-cement ratio (C.F./w/c = Omega Index Factor) in order to assess the contribution of fly ash to concrete strength.[37] He showed that for portland cement concretes having different w/c and C.F., but made with the same cements and aggregates, strengths increased linearly with the Omega Index Factor. For concretes containing fly ash, the strengths at equal Omega Index depended on a similar linear relationship. Strength differences between the concretes made with and without fly ash at equal Omega Index Factors were identified as the contribution of the fly ash to the concrete strength. It will be shown (see "Portland Cement-Fly Ash-Water Reactions" in Chapter 8) that fly ashes also affect the hydration rate of the cement, so that such differences should not necessarily be attributed to reaction of the fly ash in cases where strength increases are noted.

Tynes and Mather applied the optimization concept to very low cement content (lean) concretes for maximum economy and low heat evolution in mass concrete.[33] They showed that ample strength and watertightness for the interior of dams were attained with mixtures containing only 55.8 kg/m^3 (94 lb/yd^3) of portland cement and greater weights of fly ash. Such concrete mixtures proportioned for equal slump had decreasing water to portland cement-plus-pozzolan weight ratios with increasing pozzolan contents. Compressive strengths at 90 days increased with pozzolan contents up to a point, as shown in Fig. 1.3, beyond which strengths decreased despite lower ratios of water to cementitious material. The maximum amount of pozzolan that could be employed beneficially to contribute to compressive strength was found to decrease with increasing pozzolanic activity of the pozzolan.

Many authors have continued to develop and refine proportioning procedures for both structural and mass concrete.[38–44] ACI Standard Practice[44] for proportioning concrete containing pozzolans replaces the water-cement ratio of portland cement concrete with an equivalent water to cement-plus-pozzolan ratio. Equivalent water to cement-plus-pozzolan ratios can be calculated either on a weight equivalency or an absolute volume equivalency basis.

Dodson has recently applied the optimization concept to produce very high-strength mixtures with low portland cement and low water contents,[45] using fly ashes of high calcium contents which have cementitious as well as pozzolanic properties.

These mix proportioning methods have been used to demonstrate that with some fly ashes it is possible to proportion fly ash concretes to have strengths equal to or exceeding those of portland cement concretes at any age. However, additional factors must be considered in practice:[29]

1. Knowledge of each fly ash used must be developed to aid trial mix proportioning.
2. Concretes made with fly ash are more sensitive to temperature and moisture conditions during curing.
3. Estimates of the economic advantages of fly ash must include both cost of handling and extra components.

Properties of Fresh Concrete

The effects of fly ash on freshly mixed concrete have been summarized by Lane.[46] The reduction of the water requirement for a given consistency of concrete containing fly ash is usually attributed to the fact that most of the material is composed of spherical particles with smooth surfaces (See Fig. 3.1). In this respect fly ashes differ from other pozzolans, which usually increase the

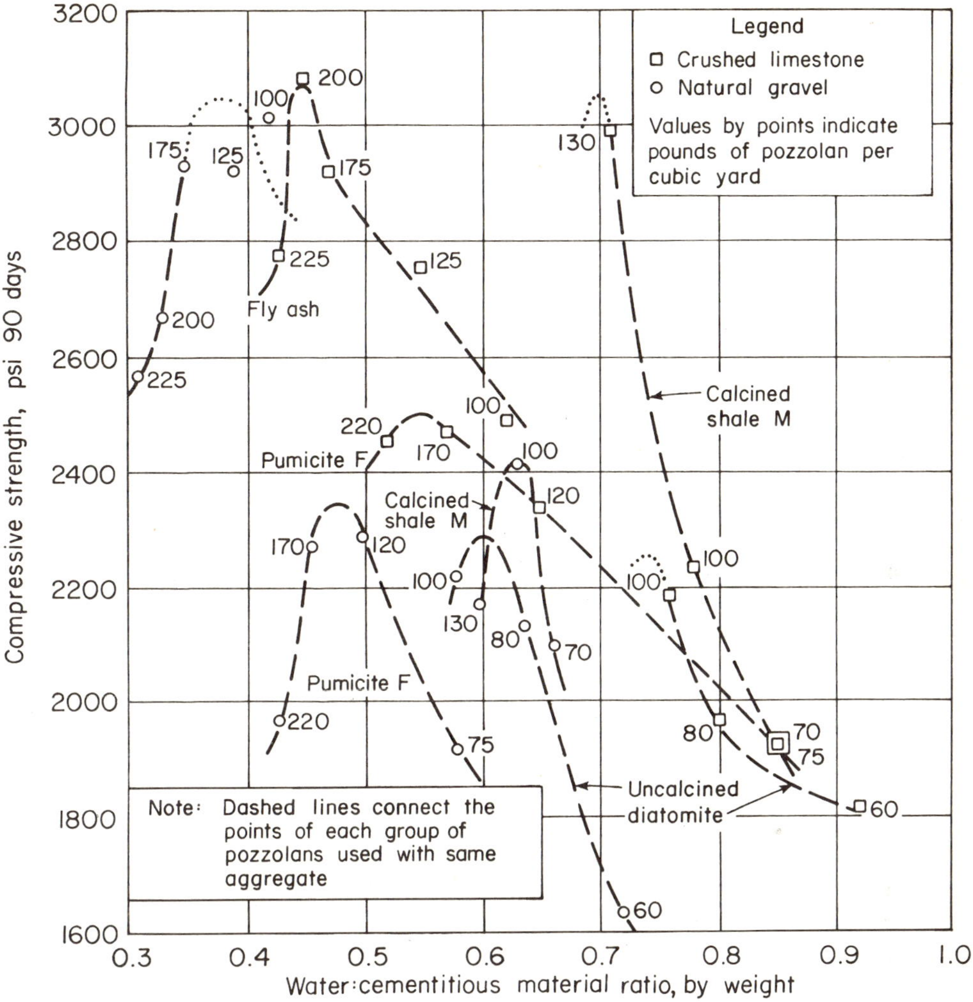

1000 psi = 6.894 MPa

Fig. 1.3 Dependence of compressive strength on water: cementitious material ratio for concrete mixtures with 55.8 kg/m³ (94 lb/yd³) of portland cement. (Ref. 33)

water requirement. Water reductions of 7% or more have been reported, but some fly ashes (usually with higher carbon content) increase water requirement. Concrete containing fly ash is generally reported to pump more satisfactorily and to show reduced segregation of aggregates and reduced bleeding of the mixing water.

The setting time of concrete is usually increased when fly ash is included, because the water to portland cement ratio is increased (i.e., the cement factor is decreased). Dodson[47] has shown that for concretes made with different water and cement contents, but with the same cement and aggregate, both initial and final setting times decrease linearly with the ratio of the cement factor, C.F. (kg/m^3), to the water to cement ratio, w/c (the Omega Index Factor). When the concretes are redesigned with fly ash replacements or additions, their setting times are equal to those of the concretes without fly ash with the same values of Omega Index Factors if only physical factors are involved. Some fly ashes chemically accelerate or retard setting. Dodson attributes the amount of the chemical effect to the deviation from the straight line on the plot of the setting time versus Omega Index Factor.

In a report containing extensive data for concretes containing Class C[(22B)] fly ashes, Majko and Pistilli[(43b)] found setting times extended by 4 to 6 hours for fly ash mixtures containing ASTM Type A water-reducing admixtures[(22B)] and very high fly ash contents.

Effect on Temperature Rise at Early Ages

The heat liberated during hydration of portland cement may total as much as 500 kJ/kg (120 cal/g) so that in structural concretes a temperature rise of 75° C might occur if no heat were lost. In mass concrete the interior may hydrate for weeks in an essentially adiabatic process and harden at elevated temperatures.[(30)] When it eventually cools, tensile stresses due to restraint of contraction develop; a temperature drop of more than 25° C must be avoided to prevent cracking.[(48)] When fly ash is used as a partial replacement of portland cement, the heat of hydration and the temperature rise of the concrete are reduced. This fact must also be recognized when making comparisons between small test specimens of portland cement and fly ash concretes to predict the strength of concrete in structures.[(49)] Fly ash concretes perform better in massive sections where the temperature rise accelerates pozzolanic reactions than is indicated by strength tests of small specimens cured at ambient temperatures.

Use of Chemical Admixtures with Fly Ash

Chemical admixtures such as water-reducing admixtures, set-controlling admixtures, and air-entraining agents can be used successfully with many fly ashes in concrete, but in some cases the admixture requirement is greater than in portland cement concrete. Tynes[(50)] found that 12 different conventional water-reducing admixtures used in concrete containing fly ash slightly improved physical properties of the concrete. Samarin and Ryan[(51)] compared the performance of plain concrete with concrete containing fly ash with and without water-reducing admixtures. Compared to plain mixes proportioned for comparable consistency and 28-day compressive strengths, the concretes with both fly ash and lignin-based admixtures have several advantages:

1. Lower bleeding capacity in leaner mixes
2. Setting times extended by one or two hours
3. Slightly reduced modulus of elasticity at 28 days
4. Higher flexural strength at 28 days

5. Comparable or lower 90-day shrinkage
6. Higher 90-day strength

The relatively new high-range water-reducing admixtures (see "Rheological Properties of Fresh Pastes" in Chapter 5) have been used with fly ash in concrete. However, the benefits claimed for use of such admixtures in plain concrete are not as apparent in fly ash mixtures.(52) The water reductions are not as great, being only slightly improved over those attainable with conventional admixtures. Since there may be less excess water in fly ash concrete, the addition of water-reducing admixtures is less effective. Nevertheless, high-strength concrete for precast structural units has been developed using such admixtures to reduce the water to cement-plus-fly ash ratio from 0.35 to 0.28 in concretes containing a very large percentage of fly ash (37% of fly ash plus cement).(53)

Effects of Fly Ash on Durability

The resistance of concrete to damage by freezing and thawing may be lowered by the use of fly ash.(54,55) However, when comparisons are made with portland cement concretes at equal strengths and adequate air contents, there are no apparent differences in the freezing and thawing durability.(55,56) Deicer scaling resistance is somewhat lower for concretes made with Type IP cements (see "Standard Specifications and Test Methods" in this chapter) containing fly ash than for Type I cements(21B) made from the same clinker.(56) The major problem concerns the increased and variable requirement for air-entraining agent in concretes with some fly ashes; these same fly ashes often have high values for loss on ignition.(55) The amount of air-entraining agent required and the loss of entrained air during extended mixing tends to be greater in concretes made with fly ashes having high organic matter content. Air contents appear to be more stable in concretes containing Class C (see "Standard Specifications and Test Methods" in this chapter) than Class F fly ashes.(57)

The effects of fly ash in concrete on its resistance to leaching by acidic waters and its resistance to chemical attack by sea water or sulfate-bearing ground waters have been extensively studied. Fly ash often improves the resistance of concrete to chemical attack. There are two major factors influencing the resistance of concrete to chemical attack: (1) the chemical composition of the cement and other ingredients of the concrete, and (2) the permeability of the concrete.

The chemical composition of the cement is notably important in the case of sulfate resistance, and special portland cement types (e.g., ASTM Type V(21B)) of low C_3A content have been developed for high sulfate resistance. The permeability of the concrete depends very strongly on the water-cement ratio and various other factors such as fineness, C_3S content, curing time, and conditions that influence the rate of hydration of the cement. When pozzolans are used, all of these factors are changed; in fact, the chemical reactions are altered in a manner which changes the relative proportions of the hydrated phases. Most importantly, the easily leachable calcium hydroxide produced in ordinary portland cement hydration is combined to a much greater extent in the relatively insoluble calcium silicate hydrate phases. The relative rates of formation and combination of calcium hydroxide in portland and pozzolanic cement pastes are shown in Fig. 8.8. It is generally accepted that in concrete this process leads to long-term gains in watertightness, strength, and resistance to aggressive environments. Fig. 1.4 shows the results of the use of fly ash on the permeability of concretes of various cement contents.[30] Although the permeability of the portland cement concretes decreased progressively with increasing cement contents, the use of a 30% fly ash-70% portland cement mixture had a much greater effect, even at low cement contents. Improved compactibility must also contribute to this effect.

Research by the U.S. Army Corps of Engineers, however, has shown the effect of fly ash on the permeability of concrete to be variable.[58] If the concrete is allowed to dry before the cement and fly ash have reacted sufficiently, the fly ash concretes can have higher permeabilities than portland cement concretes.

Sulfate attack of portland cement concrete involves both the penetration of the concrete by sulfate solutions, and therefore the permeability of the concrete, and

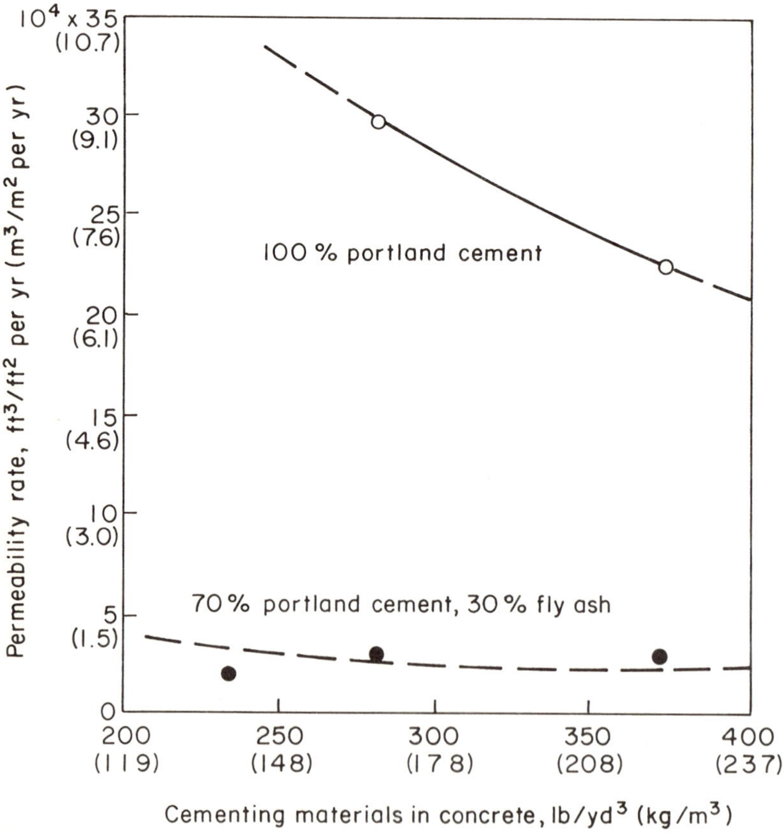

Fig. 1.4 Permeability to water of concretes made with and without fly ash. (Ref. 30)

the chemical reactions that take place between the sulfates in solution and the hydrated phases produced by portland cement.[3B,7B,59–61]

Results obtained at the U.S. Bureau of Reclamation have shown that some pozzolanic fly ashes can be very effective in reducing the expansion and deterioration caused by sulfate solutions, as shown in Fig. 1.5.[30]

Kalousek et al.[61] reported on extensive research on the long-term durability of concrete exposed to sulfate solutions. They reached the following conclusions:

1. Eighty-four percent of Type V and Type II[21B] cement concretes without pozzolans showed a life expectancy of less than 50 years.
2. Certain pozzolans increased very significantly the life expectancy of concrete exposed to 2.1% sodium sulfate solution. Fly ashes meeting present day (1972) specifications were prominent among the group of pozzolans showing the largest improvements.
3. Concretes for long-term survival in a sulfate environment should be made with high quality pozzolans and a sulfate-resisting cement. The pozzolan should not significantly increase, but preferably decrease, the amount of water required.
4. Cement to be used in making sulfate-resisting concrete with pozzolan of proven performance should have a maximum C_3A content of 6.5% and max-

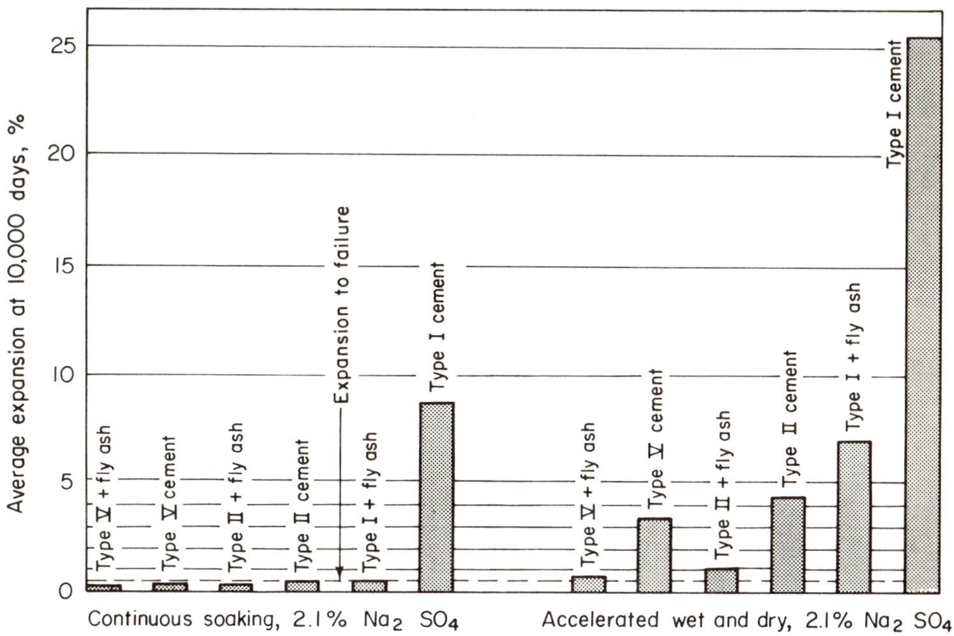

Fig. 1.5 Expansion of concretes caused by exposure to sodium sulfate solutions. Fly ash replacements were 30% of the cements. (Ref. 30) Cf. Fig. 11.1, p. 153.

imum C_4AF content of 12%. Restriction of cements to those meeting present day (1972) specifications for Type V cement[21B] does not appear justified.

These conclusions were based on results obtained with fly ashes from bituminous coals. More recent research shows that concretes made with fly ashes from lignite and subbituminous coal have generally exhibited reduced resistance to sulfate attack.[27] Attempts are being made to identify those characteristics of the fly ash that lead to reduced sulfate resistance.[27,62a] (See "Sulfate Resistance" in Chapter 11.)

The attack of sea water on concrete has some similarities to the attack of sulfate-bearing ground waters, but major differences include the relative variability of ground water pH and the variable concentrations of ions in solution, especially sodium, magnesium, and sulfate. Wetting and drying cycles also differ appreciably. Pozzolanic cements have been recommended for many years for concrete exposed to sea water, but the mechanisms of sea water and sulfate attack are not yet fully understood, and there is a need for test methods other than actual long-term exposure. Hence, it is hardly possible to specify with certainty the characteristics of fly ashes required for such purposes.

Another aspect of the durability of concrete construction is the reactivity of concrete aggregates, especially certain deleterious alkali-aggregate reactions. These reactions are not a matter of external chemical attack on the concrete but are internal reactions between cement and aggregate constituents in moist environments. At least two major categories of such reactions have been identified: alkali-silica reactions and alkali-carbonate rock reactions.[3B] In recent years certain other rock types (e.g., graywacke) have also been shown to react deleteriously.[63] Alkali-silica reactions occur most commonly with aggregates containing opaline silica such as occurs in chert, flint, shale, sandstone, limestone, and other rocks. The use of low-alkali cement (less than 0.6% total alkalies as Na_2O) has usually been effective in preventing deleterious reactions in portland cement concrete. A more flexible approach has been to limit the total alkalies per unit volume of concrete.[64]

The ability to control deleterious alkali-silica expansions with some pozzolans, including some fly ashes, has been known for some time[3B,2,3] and is recognized in ASTM standards. The effectiveness of pozzolans in controlling the expansion varies greatly, and it is necessary to test pozzolans individually.[2] As described in "Standard Specifications and Test Methods" in this chapter, test methods of ASTM C441 and C227[22B] are used for this purpose. There are, however, many questions concerning the validity of such tests (see "Alkali-Aggregate Reaction Control" in Chapter 11).

Control of alkali-aggregate reactions by the use of pozzolans will tend to become even more important in the future as the best sources of concrete aggregates are depleted and as energy conserving processes in cement manufacture become general practice. These processes retain more of the alkalies in the cements than do less efficient processes. However, deleterious alkali-carbonate rock reactions have been shown to be relatively unaffected by pozzolans.[65]

Corrosion of reinforcing steel can be a serious problem in structures exposed to marine environments, bridge decks exposed to chloride-containing deicing chemicals, and areas such as the Middle East, where the materials used and the severe environment combine to accelerate corrosion processes.[66,67] Steel reinforcement is ordinarily well protected against corrosion when it is encased in sound, well-compacted, portland cement concrete.[7B] Laboratory studies have shown that corrosion protection in aggressive environments can be increased by inclusion of fly ash in concrete when comparisons are made between concretes of equal 28-day strengths.[68]

Suspicions that steel reinforcement in fly ash concrete might be more susceptible to corrosion because of the reduction of the pH of the aqueous solutions by the pozzolanic reactions do not appear to be well founded.[29,68] Drying and neutralization by carbonation (reaction of cement hydrates with atmospheric carbon dioxide) are of greater concern, because corrosion begins when the depth of carbonation (and reduced pH) reaches the steel. The depth of carbonation has been found to be greater for concretes proportioned by replacement of cement with fly ash and exposed to drying after only short (7-day) curing periods.[69,70] However, in a comprehensive study of all properties of concretes proportioned to take advantage of the water reduction achieved with fly ashes, Lewandowski[71] found the carbonation depths to be within the range of values attained without fly ash, irrespective of preliminary curing and storage conditions.

Standard Specifications and Test Methods

The original ASTM tentative specification for fly ash adopted in 1954 was based largely on work done with fly ashes from bituminous coals; it did not distinguish between fly ashes produced from different types of coal, nor did subsequent revisions until 1977. In 1968 the specification was combined with that for raw or calcined natural pozzolans in ASTM C618-68T, which was adopted as a standard in 1971. At that time fly ashes were designated as "pozzolan Class F" to distinguish them from other pozzolanic materials (Class N and Class S). Many fly ashes obtained from lignite coal and subbituminous coals from the Western United States and Canada did not meet the 70% minimum $SiO_2 + Al_2O_3 + Fe_2O_3$ content for the Class F fly ashes because of their high calcium contents. Although they did not meet the ASTM specification, it was shown that many of these fly ashes were suitable for use in concrete.[72] In 1977 two classes of fly ashes were recognized in ASTM C618: Class F, fly ash normally produced from burning anthracite or bituminous coal, and Class C, fly ash normally produced from lignite or subbituminous coal. Class F was recognized as having pozzolanic properties; Class C fly ash was recognized as having some cementitious properties as well, and it was reported that "some may contain lime contents higher than 10%." Manz[72] has termed the ashes from subbituminous and lignite sources "basic ashes." He has also noted that these lignite ashes may have high contents of free CaO in a dead-burned state. Dead-burned free CaO in portland cement is well known to be a cause of unsoundness, and excessive amounts in concrete must be prevented to ensure the soundness of the concrete. A summary of the

1986 chemical and physical requirements for fly ashes in ASTM C618 is given in Table 1.1.[1]

ASTM Tentative Methods of Sampling and Testing Fly Ash for Use as an Admixture in Portland Cement Concrete were issued in 1953 and revised until 1968, when they were adopted as Standard Methods. The current standard

Table 1.1 Summary of 1986 Chemical and Physical Requirements Defined in ASTM C618-85 for Fly Ashes for Use in Portland Cement Concrete

	Class F	Class C
A. Chemical requirements		
(SiO_3 + Al_2O_3 + Fe_2O_3), min.,%	70.0	50.0
SO_3, max., %	5.0	5.0
Moisture content, max., %	3.0	3.0
Loss on ignition, max., %	6.0*	6.0
Optional chemical requirements		
Available alkalies, as Na_2O, max., %	1.50	1.50
B. Physical requirements		
Fineness:		
Amount retained when wet-sieved on No. 325 (45 μm) sieve, max., %	34	34
Pozzolanic activity index:		
With portland cement, at 28 d, min., % of control	75	75
With lime, at 7 d, min., psi (kPa)	800(5500)	
Water requirement, max., % of control	105	105
Soundness:		
Autoclave expansion or contraction, max., %	0.8	0.8
Uniformity:		
Specific gravity, max. variation from average, %**	5	5
Percent retained on No. 325 (45 μm) sieve, max. variation, percentage points from average**	5	5
Optional physical requirements		
Multiple factor, product of LOI, %, and fineness, %, max.	255	—
Increase of drying shrinkage of mortar bars at 28 d, max., %	0.03	0.03
Uniformity requirements:		
When air-entraining concrete is specified, quantity of AE agent required to produce air content of 18.0 vol % shall not vary from the average of previous 10 by more than, %	20	20
Reactivity with cement alkalies:		
Mortar expansion at 14 d, max., %	0.020	0.020

*Reduced from 12.0% in 1984, with the provision that Class F pozzolan containing up to 12.0% loss on ignition may be approved by the user if either acceptable performance records or laboratory test results are made available.

**Average established by the ten preceding tests, or by all the preceding tests if the number is less than ten.

Table 1.2 Standard Tests of Fly Ashes Contained in ASTM C311-85

Chemical Tests	Physical Tests
Moisture content	Specific gravity
Loss on ignition	Fineness
Silicon dioxide	Increase of drying shrinkage of mortar bars
Aluminum oxide and iron oxide	Soundness
Calicum oxide	Limits on amount of air-entraining admixture in concrete
Magnesium oxide	Air entrainment of mortar
Sulfur trioxide	Pozzolanic activity index with portland cement
Available alkalies	Water requirement
	Pozzolanic activity index with lime
	Reactivity with cement alkalies

ASTM C311-85[73] includes methods for chemical analysis and physical tests, as indicated in Table 1.2.

The soundness requirement in Table 1.1 is to guard against expansive reactions caused by the presence in the fly ash of excessive amounts of free CaO, periclase, or sulfate. The autoclave expansion test, ASTM C151,[21B] modified as specified in ASTM C311 (paste bars made with pozzolan to cement ratios of 25:100) is the procedure used for the evaluation of the effect of a fly ash on the soundness of concrete in these respects. The requirements in ASTM C618 include only two for uniformity: specific gravity and fineness by the residue on a No. 325 sieve. ASTM C311 specifies that tests for fineness, moisture content, specific gravity, loss on ignition, and soundness be made on each 400 Mg (400 tons) of fly ash. Chemical analysis and other tests are made on composite samples of each 2000 Mg (2000 tons). These tests are not very frequent when it is considered that a mile of concrete highway pavement may require only 180 to 270 Mg of fly ash (200 to 300 tons).

The Multiple Factor was introduced to limit the effects of large amounts of carbon and very coarse material in any one ash on the water requirements of mortars and concretes (see "Water Requirements of Mortars and Concretes" in Chapter 5). The Multiple Factor is the product of the residue on the No. 325 sieve and the loss on ignition. It was adopted as an optional requirement by ASTM in C618 in 1971 and is limited to a maximum of 255 for Class F fly ashes. It was not adopted for Class C fly ashes because the lower LOI limitation predominates in those ashes.

The mortar expansion in the optional requirements in Table 1.1 is determined in accordance with the requirements of ASTM C441[22B] Standard Test Method for Effectiveness of Mineral Admixtures in Preventing Excessive Expansion of

Concrete due to the Alkali-Aggregate Reaction (see "Alkali-Aggregate Reaction Control" in Chapter 11). The fly ash is tested in the "job mixture" of fly ash and cement with crushed Pyrex glass of specified grading. The tests are made by the methods of ASTM C 227, Standard Test Method for Potential Alkali Reactivity of Cement Aggregate Combinations (Mortar Bar Method).[22B]

A Standard Specification for Blended Hydraulic Cements (ASTM C 595-68) was adopted in 1968. This restricted the pozzolan contents to a range of 15 to 40% pozzolan in portland-pozzolan cements. The specification included chemical, physical, and performance test limits for portland-pozzolan cements and only two physical requirements for pozzolans (20.0% maximum retained on a No. 325 sieve and 800 psi (5.5 MPa) in the pozzolanic activity strength test). In 1979 a new category of pozzolan-modified portland cement with less than 15% pozzolan was designated as Type I(PM) in ASTM C 595. In the 1986 ASTM specifications,[74] portland-pozzolan cements are classified into three major categories according to use:

Type P	Portland-pozzolan cement for use in concrete construction where high strengths at early ages are not required.
Type IP	Portland-pozzolan cement for use in general concrete construction.
Type I(PM)	Pozzolan-modified portland cement for use in general concrete construction.

Each of these major categories may be subdivided into eight subclassifications. Moderate sulfate resistance and air entrainment, or both, may be specified by adding the suffixes (MS) or (A). In addition, moderate heat of hydration [for Types IP and I(PM)] or low heat of hydration (for Type P) may be specified by adding the suffixes (MH) or (LH). A summary of the 1986 requirements for pozzolans in ASTM C 595-86 is given in Table 1.3.[74] The only specific chemical requirements for pozzolans are those implied by the maximum amounts of MgO (5.0%), SO_3 (4.0%), and loss on ignition (5.0%) specified for the entire portland-pozzolan cement composition. To detect possible deleterious (alkali-aggregate type) expansive reactions, six additional alkali reactivity expansion tests are specified in Table 1.3 for pozzolans that are to be used in pozzolan-modified portland cements, Type I(PM). These tests are performed by ASTM C 227 using

Table 1.3 Summary of 1986 Physical Requirements Defined in ASTM C595-86 for Fly Ash in Blended Cement

Fineness:	
Amount retained when wet sieved on 45 μm (No. 325) sieve, maximum %	20.0
Pozzolanic activity index:	
With portland cement, at 28 days, minimum % of control	75.0
Alkali reactivity of pozzolan for use in Types I (PM) and I (PM)-A cement:	
Mortar bar expansion at 91 days, maximum % (6 tests with cements containing 2.5, 5, 7.5, 10, 12.5, and 15% pozzolan)	0.05

a nonreactive aggregate. If the blended cement is to be used with an alkali-reactive aggregate, an additional optional requirement may be specified. In such cases the mortar expansion by C227 (using crushed Pyrex glass of a specified grading as aggregate) is limited to 0.02% at 14 days and 0.06% at 8 weeks.

National standards of many countries for fly ashes have been compared by several authors.(75–78) Some fly ashes with high free CaO contents will produce rapid heat evolution, temperature rise, and false set in concrete mixes. It is thought likely that limits for free CaO may be set in ASTM specifications in the future(75) to control the use of such fly ashes. For several years ASTM Subcommittee C01.29 on Sulfate Resistance has been attempting to develop a test for sulfate resistance for hydraulic cements including blended cements and blends of portland cements with fly ashes(79a) (see "Sulfate Resistance" in Chapter 11). In 1984 a Standard Test Method for Length Change of Hydraulic-Cement Mortars Exposed to a Mixed Sodium and Magnesium Sulfate Solution (ASTM C 1012-84) was adopted.(79b)

Several authors have advocated specific quality control measures for fly ashes.(42,80–86) Two new rapid tests have been suggested to monitor variability of Class C fly ashes.(80,81) One is the "Self-Setting" test in which 10 g ash is mixed with 30 ml water, formed into a pat, and tested with a spatula for the time required for it to set enough to permit a clean cut through the pat. Specimens that have not set in 3 hours at 73° F (23° C) are considered nonhydraulic.

The second test is the "Acid Temperature Rise" test, which is purported to be a measure of total CaO content, determined by measuring the temperature rise of 20 g of fly ash in 75 ml of 15% HCl in a vacuum flask. The residue may be washed and dried to determine acid solubles.

The Need for Improved Characterization

Several authors have critically reexamined and discussed the evolution of ASTM specifications and test methods.(10,75,77–81) Some authors have advocated more restrictive chemical and physical specifications to assure concrete performance. Others have sought to supplant such methods with performance tests and specifications to the maximum possible extent, while recognizing the inadequacy of our knowledge and many of our test methods.

The chemical composition limits specified in ASTM C618 (Table 1.1) for Class F and Class C fly ashes have been criticized for not differentiating between different fly ashes as intended. These classifications are based on the distinction between ashes with only pozzolanic properties (Class F) and those which also have some cementitious properties (Class C). Cementitiousness depends primarily upon the calcium oxide content so that calcium silicate and aluminate hydrates may be formed. However, C618 contains no direct specification for the total CaO content, nor is there any specific test for cementitiousness or any difference in the pozzolanic activity index tests or specification for the two classes of ash. Instead, the only major specification difference is the minimum limit of $SiO_2 + Al_2O_3 + Fe_2O_3$, which is 70% for Class F fly ash and 50% for Class C. This difference indirectly recognizes that the cementitious ashes contain appreciable

CaO by allowing a lesser sum of silica, alumina, and iron oxides than for the Class F ashes. Lignite and subbituminous coals usually have high-calcium content ashes but these, and even bituminous coals, can vary widely in CaO contents, as demonstrated by the ranges shown in Table 2.3.

Revised criteria for classification of fly ashes have been proposed by Kocuvan,[87] Mielenz,[82] and Nicolescu[88] in 1979, by Cain[89] and Manz[75] in 1981, and by Butler[80] and Diamond[90] in 1982. Diamond[90] has proposed that the chemical specifications require determination of the total CaO contents and that, instead of Class C and Class F, fly ashes be classified into three groups with subdivisions at 5 and 15% CaO.

The need for further research for improved characterization of fly ashes for use in concrete is widely recognized, but there is less agreement on what needs to be done.[50B] There is a special need for rapid test methods that can be used with confidence to predict long-term performance.[81,85,91] This predictive capability will probably require the development of accurate mathematical models or artificial intelligence systems that can also be used to establish improved standard specifications in terms of materials characteristics and formulations.

In the next chapter we begin to examine the state of our knowledge of fly ash with a description of its generation in coal-fired power plants. Because the properties of fly ashes depend strongly on the particular coal used, we begin with an examination of the kinds of coal.

Chapter 2

Generation of Fly Ash

The structures, compositions, and properties of fly ash particles depend upon the structure and composition of the coal and the combustion processes by which fly ash is formed. The fly ashes of the two major classes (F and C) recognized by ASTM for use in concrete are derived from different kinds of coal.[1] The different kinds of coal result from differences in the coal formation processes over geological time periods.

Coal

The Origin of Coal

Coal is formed by the decomposition of plant matter, without free access to air, under the influence of moisture, pressure, and temperature.[92,25B] During the first (biochemical) stage of coal formation, complete decay of plant material is prevented only in regions, such as swamps, where there is plentiful plant growth. The submerged accumulation of plant debris forms peat in the first step of the coalification process. During this biochemical stage, different plant parts are digested by fungi and bacteria at different rates. The most decay-resistant parts include the waxy protective layers in cuticles, spores, and pollen, as well as the resins.

The geochemical stage of the conversion of peat to bituminous coal begins with the covering of peat with sediment during repeated subsidence and submergence of peat swamps, resulting in interbedded layers of silt, sand, clay, and peat. The covering of sediment provides pressure for compaction and thermal insulation, so that compression at elevated temperatures occurs. Additional pressure and temperature increases generated by mountain formation processes can result in anthracite or, in some cases, a natural coke.

Most bituminous coal fields were deposited during the Carboniferous and upper Cretaceous Periods, 300 to 100 million years ago. Coals in which the degree of coalification is less complete (subbituminous, brown coal, and lignite) were formed more recently. Peat deposits are generally less than one million years old.

Kinds of Coal

Coals are classified by type, rank, and grade, as described in the following subsections.

Coal Types and Petrography Different *types* of coal result from variations in the amounts of different plant parts preserved in the coal. These variations depend on the degree of decomposition in the biochemical stage. The most common banded coal types are known as bright coal, semisplint, and splint (dull) coal. Nonbanded coals such as cannel coal and boghead coal are believed to have formed in aerated water which decomposed all of the woody plant tissues. It is believed that cannel coal was formed from spores and pollen and that boghead coal was derived from algae. Fig. 2.1 shows examples of banded and nonbanded coals.[25B]

Several overlapping systems of terminology and classification of coal types have developed, based on macroscopic observation and microscopic examination of thin sections or polished samples.[92,25B–27B] Examinations made with transmitted light through thin sections permit the identification of the plant parts that became coal components. By this method three components are identified: anthraxylon, attritus, and fusain. These components are the basis of the

Fig. 2.1
Top: Banded bituminous coal. No. 6 bed, Franklin County, Illinois. Shows alternations of vitreous bright coal (vitrain) with thinly laminated coal (clarain). U.S. Bureau of Mines photograph. (Ref. 25B)

Bottom: Nonbanded cannel coal. Shows blocky structure, conchoidal fraction, and absence of stratification. U.S. Bureau of Mines photograph. (Ref. 25B)

Theissen-Bureau of Mines classification system. The proportions of these components determine the type of coal.

Anthraxylon is translucent matter derived chiefly from the woody parts of plants; the thicker bands were large branches or trunks of trees, and the thinner bands were twigs or small branches. Attritus was formed from tightly compacted, macerated, and finely divided plant materials. It is a dull gray to nearly black composite. Translucent attritus includes humic matter similar to anthraxylon but with particle sizes smaller than 14 μm; the opaque attritus contains a higher proportion of granular opaque matter and black amorphous material similar to finely divided (less than 37 μm) fusain.

Fusain, or mineral charcoal, is characterized by its silky luster, fibrous structure, and black color. The relict plant cell structure can be observed. If it is not mineralized, it is very friable and highly porous. The pores are sometimes filled with calcite ($CaCO_3$) or with pyrite (FeS_2).

In terms of these microscopically determined constituents, the main coal types are defined as follows:[92]

Banded Coals	
bright coal	more than 5% anthraxylon less than 20% opaque attritus
semisplint coal	more than 5% anthraxylon 20 to 30% opaque attritus
splint (dull) coal	more than 5% anthraxylon more than 30% opaque attritus
Nonbanded Coals	
cannel coal	less than 5% anthraxylon
boghead coal	less than 5% anthraxylon

Specimens of banded coals under ordinary visual observation appear to be composed of four varieties. This classification system uses four rock types designated as vitrain, clarain, durain, and fusain.[26B]

Vitrain (primarily anthraxylon) is a coal lithotype characterized by a brilliant, vitreous luster, black color, cubical cleavage, and conchoidal fracture. Vitrain bands or lenticles are amorphous, usually 3 to 5 mm (up to 20 mm) thick.

Clarain is characterized by a semibright, silky luster and sheetlike irregular fracture. It is distinguished from vitrain by its fine intercalations of durain.

Durain (mainly opaque attritus) is characterized by a dull, matte luster, gray to brownish black color, and granular fracture. It occurs in bands up to many centimeters in thickness.

A third system of terminology (Stopes-Heerlen) for description and classification of coals is based on microscopic observation of polished surfaces using reflected light. In this system, the relicts of individual plant parts (macerals) are regarded as the counterparts of the minerals in inorganic rocks. The macerals may be grouped into three major maceral groups: vitrinite, exinite, and inertinite, which roughly correspond to the anthraxylon (vitrain), translucent attritus, and opaque attritus of the Thiessen-Bureau of Mines system.[92] The Stopes-Heerlen system then recognizes eight microlithotypes, which are classified according to the relative amounts of macerals present in typical samples. The terminology is rarely encountered in the present literature on fly ash.

Ranks of Coal Different *ranks* of coals refer to their degree of coalification. The chemical composition of the coal gradually changes during coalification to compounds of lower oxygen and hydrogen content and higher carbon content in aromatic ring structures. The percentage of volatile matter decreases and the calorific value increases. These latter quantities are the basis for classifications of coals by rank through a continuous series: peat, brown coal and lignite, subbituminous coal, bituminous coal (soft coal), and anthracite (hard coal).[92,93] The end member of this series is graphite.

Lignite and brown coal are in the lowest rank of the solid fuels known as coal. Lignite is the term applied to the consolidated coal. It has a high moisture content and checks badly when dried. It is composed of metamorphosed woody and other plant tissues, which are usually easily recognizable. Brown coal is an unconsolidated friable material.

Subbituminous coal is a black and crumbly coal intermediate between lignite and bituminous coal.

Bituminous coal (soft coal) is the most abundant rank of coal. It is usually banded, dark brown to black in color, and does not disintegrate on exposure as readily as subbituminous coals.

Anthracite (hard coal) is the highest rank of coal. Due to metamorphism, the fixed-carbon content of anthracite is between 92 and 98% on the dry weight basis. Hard and black, it has a semimetallic luster and semiconchoidal fracture.

Analyses of coal customarily are given in the form of proximate constituents, which include moisture, volatile hydrocarbons, fixed carbon, and ash.[94] Total moisture is determined by heating to constant weight at 107° C. Volatile matter is determined by weight loss at 950° C under conditions that allow a char of fixed carbon and mineral matter to be formed. Ash is determined by ignition at 750° C and fixed carbon by difference. Detailed ASTM specifications for the different ranks of coal are given in Table 2.1.

Grades of Coal The inorganic constituents in coal cause differences in *grade*. Carbon-containing deposits such as oil shale and bituminous shale, in which inorganic materials predominate, are not classified as coal. The average ash content of coal used in steam-generating plants is about 15% in the United States, the normal range being from 3 to 30%. The ash content depends on methods of mining and cleaning the coal, as well as on the inherent ash content. The use of continuous mining equipment tends to include more rock adjacent to the coal seams; such run-of-mine coal currently includes about 25% ash.[92]

Composition of Coal

Both the inorganic and organic constituents in coal affect the composition of the fly ash produced by combustion. The ash consists mainly of inorganic glassy

Table 2.1 ASTM D 388-84 Classification of Coals by Rank[a]

(Ref. 93)

Class	Group	Fixed carbon limits, percent (dry, mineral-matter-free basis): Equal or greater than	Fixed carbon limits: Less than	Volatile matter limits, percent (dry, mineral-matter-free basis): Greater than	Volatile matter limits: Equal or less than	Calorific value limits, Btu per pound (moist,[b] mineral-matter-free basis): Equal or greater than	Calorific value limits: Less than	Agglomerating character
I. Anthracitic	1. Meta-anthracite	98	—	—	2	d	—	Nonagglomerating
	2. Anthracite	92	98	2	8	d	—	
	3. Semianthracite[c]	86	92	8	14	d	—	
II. Bituminous	1. Low-volatile bituminous coal	78	86	14	22	d	—	Commonly agglomerating[e]
	2. Medium-volatile bituminous coal	69	78	22	31	d	—	
	3. High-volatile A bituminous coal	—	69	31	—	14,000[d]	—	
	4. High-volatile B bituminous coal	—	—	—	—	13,000[d]	14,000	
		—	—	—	—	11,500	13,000	
	5. High-volatile C bituminous coal	—	—	—	—	10,500	11,500	Agglomerating
III. Subbituminous	1. Subbituminous A coal	—	—	—	—	10,500	11,500	Nonagglomerating
	2. Subbituminous B coal	—	—	—	—	9,500	10,500	
	3. Subbituminous C coal	—	—	—	—	8,300	9,500	
IV. Lignitic	1. Lignite A	—	—	—	—	6,300	8,300	Nonagglomerating
	2. Lignite B	—	—	—	—	—	6,300	

[a]This classification is applicable to coals that are composed mainly of vitrinite. Certain coals, principally nonbanded varieties, rich in inertinite or exinite do not fit into the classification.

[b]Moist refers to coal containing its natural inherent moisture but not including visible water on the surface of the coal.

[c]If agglomerating, classify in low-volatile group of the bituminous class.

[d]Coals having 69% or more fixed carbon on the dry, mineral-matter-free basis shall be classified according to fixed carbon, regardless of calorific value.

[e]It is recognized that there may be nonagglomerating varieties in these groups of the bituminous class and that there are notable exceptions in high-volatile C bituminous group.

1 Btu/lb = 2326 J/kg

particles formed from the mineral matter in the coal.

The mineral matter in coal is the result of at least three different processes: (1) growth of the original plant tissue, (2) deposition of (detrital) minerals during and after peat formation, and (3) crystallization of secondary minerals from water which permeated the coal seams. Among the most common detrital minerals are the clay minerals (various finely divided crystalline aluminum, iron, and magnesium silicates). Quartz, feldspars, and garnets are also common. The most common secondary minerals are kaolinite, calcite, and pyrite. Some Western U.S. coals have been characterized as having high calcium (as gypsum) and low pyrite contents compared with Midwestern and Eastern U.S. coals.(95) Murtha(96) has compiled a complete listing of power plants expected to be using western coals by 1985, including the mine sources and tonnages.

The mineral matter in coal is more or less uniformly distributed as small inclusions of approximately 2 μm mean size or segregated in layers or agglomerates. However, much larger (20 to 60 μm) pyritic crystals and other particles have been observed.

Coal is composed mainly of carbon, hydrogen, and oxygen. Nitrogen and sulfur are lesser constituents. Both ultimate (elemental) and proximate analyses of several coals of different ranks are given in Table 2.2.(97) The ash consists mainly (95 to 99%) of oxides of Si, Al, Fe, and Ca; smaller amounts (0.5 to 3.5%) of Mg, Ti, S, Na, K, and trace amounts of many elements are also present.(33B)

Low sulfur content is required to meet EPA air quality standards. To meet the present standards (490 g $SO_2/10^6$ Btu or 460 g SO_2/GJ), steam coals must contain less than 1% sulfur, unless the boilers are used with scrubbing equipment to remove SO_2.(33B) Most sulfur in high-sulfur coals is present in mineral impurities, but in very low-sulfur coals the organic sulfur exceeds the inorganic. Many other elements are present in coals in only trace amounts. Some of these, such as lead and mercury, are toxic and are of considerable importance with respect to air pollution during combustion or with respect to safe utilization or disposal of fly ash.(14,33B,35B)

The most important oxygen-containing functional groups in coals are carbonyl, hydroxyl, carboxylic acid, and methoxy. The main nitrogen-containing groups are aromatic nitriles, pyridines, and pyrroles. Organic sulfur is found mostly in thiophene groups; other forms in which sulfur occurs are aromatic mercaptans, disulfides, and sulfides.

Studies of vitrinites in bituminous coals and anthracite provide a good conception of the structure of the coal "molecule." The molecule consists of a number (dependent on coal rank) of aromatic nuclei or clusters, each cluster usually containing one to four fused benzene rings (naphthalene, anthracene, etc.). These aromatic clusters are partly linked together by both short groupings (such as methylene, ethylene, or methoxy) and other ring structures. A small fraction of the ring structures may be heterocyclic, i.e., may have an oxygen (or a nitrogen or sulfur) atom substituted for a carbon atom; some of these may be five-membered rings such as thiophene. The molecules are composed of planar-fused ring structures linked to nonplanar hydroaromatic structures, as indicated in the proposed model shown in Fig. 2.2. This structure is irregular, open, and com-

Table 2.2 Typical Proximate and Ultimate Analyses for Coals and Char[a]

(Table from Ref. 97)

Coal I.D. and rank	Utah Church Mine, bituminous	Pittsburgh,[b] bituminous	Pittsburgh,[b] bituminious (high volatile)	Sewell[b] bituminous (medium volatile)	Anthracite[b] (low volatile)	Illinois coal, bituminous	Illinois[c] coal char. —
Moisture, %	2.5-2.7	2.0	1.0	1.9	1.3	10.1	0.9
Proximate, %							
Volatiles	44.1-45.5	36.6	28.9	16.3	8.8	39.9	2.4
Fixed carbon	42.6-44.2	55.4	63.2	75.6	71.8	52.0	76.8
Ash	9.2-9.5	6.0	6.9	6.2	18.1	8.1	20.8
Ultimate, %							
Carbon	69.8-71.5	77.5	80.6	84.2	73.2	68.3	74.0
Hydrogen	5.5-5.6	5.3	4.9	4.3	3.1	5.0	0.7
Nitrogen	1.4-1.5	1.5	1.5	1.2	0.9	1.3	1.0
Sulfur	0.4-0.7	1.2	0.7	0.7	0.9	3.5	3.3
Oxygen	11.2-13.2	8.5	5.4	3.4	3.8	13.8	0.2
Ash	9.2-9.5	6.0	6.9	6.2	18.1	8.1	20.8

[a] Data courtesy of Bureau of Mines, Brigham Young University, and ERDA.
[b] From U.S. Bureau of Mines, Pittsburgh, Pa.
[c] From COED Gasification, supplied by Energy Research and Development Administration, Pittsburgh, Pa. (4th stage product).

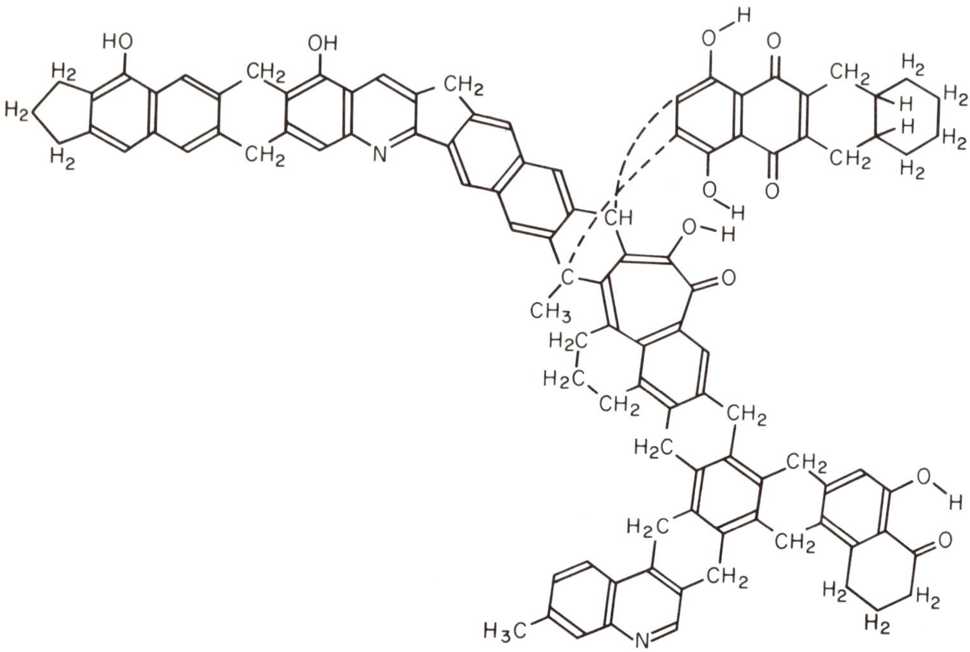

Fig. 2.2 Model of the coal molecule. (Ref. 97)

plex, with an extensive ultrafine porosity. In bituminous coals, a tarlike material partially fills the interstices between the molecules, and most of the moisture is contained in these capillaries. Vapors pass through these passages during thermal decomposition.

Combustion and Ash Formation

Most fly ash is produced by utility boilers burning pulverized coal. Stoker-fired units produce relatively coarse fly ash. The new fluidized bed combustion systems also produce coarse fly ash.[15] For pulverized coal-fired boilers, the coal is ground in a coal mill to high fineness, frequently 70% finer than 74 μm (No. 200-mesh sieve). The mass median diameter is then about 50 μm. The amount of ash may be reduced by washing and drying the coal. However, for efficient grinding, the moisture content must be low, and it may be more economical to use dry run-of-mine coal with up to 30% ash than to clean and dry the coal.

Various firing methods are illustrated in Fig. 2.3.[43B,98] Higher rank coals are slower burning and require large furnaces to complete combustion. Lower rank coals are better suited for cyclone burners which carry out rapid, intense combustion. Cyclone burners typically yield only 10% fly ash and 90% bottom ash. Most other pulverized coal-firing methods yield 65% fly ash and 35% bottom ash (as slag in wet bottom boilers) or 85% fly ash and 15% bottom ash (dry bottom boilers). The bottom ash particles are much coarser than the fly ash.

A model of the process of fly ash formation presented by Flagan[99] is illustrated in Fig. 2.4. Malte and Rees[100] described the process as follows:

> The model begins with the heating and devolatilization of the coal particles and the initial char burnout. For most pulverized-coal particles, this results in a fragile, skeletal char structure which continues to support the mineral inclusions. The exception is coal particles that become plastic and swell upon rapid heating (for example, vitrite particles). These tend to form hollow vesicular spheres. The hot, molten ash inclusions tend to fuse together, rather than appreciably vaporize, as the char material burns away. This agglomeration occurs principally during the last 25% of char burnout. At some point during char burnout (Flagan suggests 90% char burnout), the weakened char structure fractures. The number of resultant fragments determines the number of large flyash particles formed per original coal particle. For nonswelling lignite, Padia et al.[101] indicate an average of three fragments per coal particle, while for bituminous coal, with some swelling, the yield is five fragments per coal particle.
>
> The formation of hollow spherical flyash particles with coherent nonporous shells, called cenospheres, has been the subject of considerable attention as well as uncertainty. The experimental results of Padia et al. for simulated furnace conditions, indicate substantial cenosphere formation during combustion of both pulverized lignite and bituminous coals.
>
> On the other hand, Field et al.[102] indicate negligible cenosphere formation with anthracite coals. Raask found as much as 20% by volume, or 5% by mass, of furnace flyash from British coals to be cenospheres; he indicated a size range of 20 to 200 μm, which is considerably above the size of the dense flyash particles.[103]
>
> Both Raask and Padia et al. indicate an optimum particle temperature for cenosphere formation. The temperature must be high enough so that the viscosity of the molten ash is low, yet at the same time, the temperature must be low enough so that the internal gas pressure, which facilitates cenosphere formation, is not so great as to cause the cenosphere shell to rupture. Padia et al. suggest the range 950 to 1130° C, while Raask indicates a temperature of 1400° C for cenosphere formation. Part of this difference in reported ranges could be due to experimental uncertainty in the true particle temperature, or due to differences in ash composition.
>
> The results of Padia et al. imply that a cenosphere could form directly from a small ash inclusion in the coal, or from an agglomerate due to the fusion of a few nearby inclusions. They indicate a 30-ms formation time for a 12-μm particle to form a 40-μm diameter cenosphere at 1130° C . . . .
>
> To this point, the discussion has concerned the large-sized flyash. Flyash particles of submicron size appear to be a result of ash vaporization and subsequent condensation [see Fig. 2.4]. Ash vaporization occurs in several forms: (1) above about 1080° C, alkali-metal salts vaporize; (2) at high temperatures, compounds of heavy metals such as As, Cd, Cr, Ni, Pb, Se, Sb, Tl, and Zn will boil or sublime; and (3) above 1630° C, silica volatilizes by first forming the more volatile SiO according to

$$SiO_2 + C^* \rightarrow SiO + CO \tag{2.1}$$

[where C* refers to solid carbon]

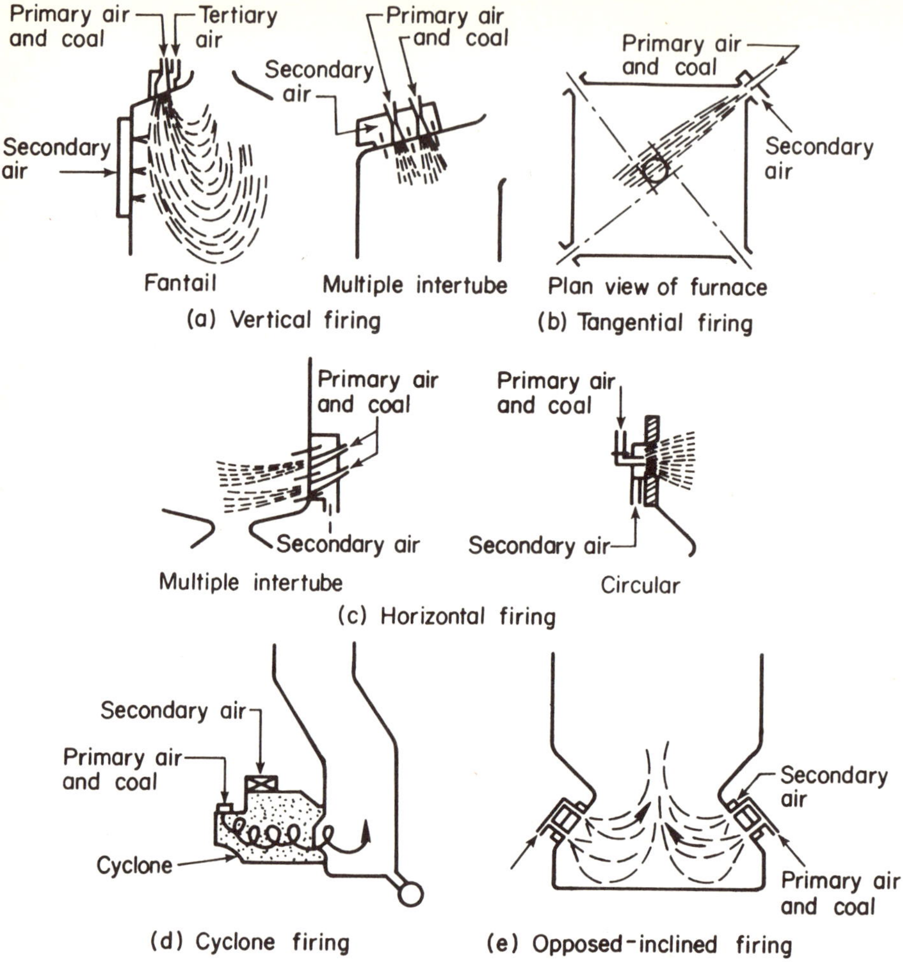

Fig. 2.3 Various types of furnaces for firing coal. (Ref. 98)

As the furnace gases cool, the inorganic vapors undergo homogeneous nucleation and heterogeneous condensation on existing particles. Because small particles have greater specific surface areas, they tend to collect larger quantities of the condensed metallics, as well as condensable sulfates.

Ulrich et al.[104,105] have proposed a mechanism to explain the growth of submicron particles. The theory relies on particle collison rates determined by free-molecule Brownian motion, followed by coalescence. The model appears to explain the measured growth in particle size for times of 30 to 100 ms following initial homogeneous nucleation.

During cooling, inorganic species in the vapor phase begin to condense as thin coatings (0.01 to 0.03 μm thick) on the larger particles and also as very fine primary particles (0.06 μm), which then grow by coagulation in the combustion zone for several seconds before being quenched.[106] These particles and the

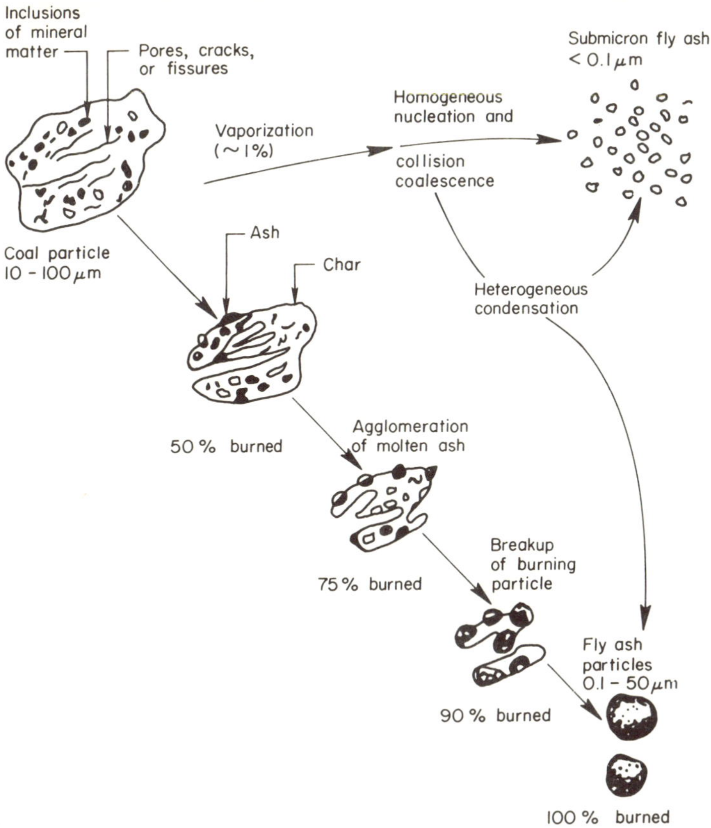

Fig. 2.4 Coal combustion and fly ash formation processes. (Ref. 99)

surfaces of the larger particles are rich in inorganic components such as alkali sulfates and potentially toxic trace elements.[107–109]

Upon leaving the combustion zone, the ash particles are cooled quickly, 1500° C to 200° C in 4 seconds being reported.[33B] Not all of the organic material is completely oxidized during combustion. Various complex organic compounds closely related to those present in coal have been identified in fly ashes. Organic residues, such as polycyclic aromatic hydrocarbons, are also adsorbed on the particle surfaces. These organic compounds may resist thermal and photochemical decomposition while adsorbed on the surfaces of the fly ash particles.[110] This picture of the process suggests that although the composition of the bulk material in the larger particles may vary widely, depending on the composition of the individual coal particles, the material deposited on the surfaces from the vapor phase is relatively uniform.

Composition of the fly ash can be affected by various substances injected into the coal or gas stream to reduce gaseous pollutants or to improve efficiency of particulate collectors. Limestone and dolomite injection processes for desulfurization of the exit gases have been investigated. If such methods were used, the calcium and magnesium contents of the fly ash would be increased.

Conditioning agents such as sulfur trioxide, sodium carbonate and bicarbonate, sodium sulfate, phosphorus, magnesium oxide, water, ammonia, and triethylamine are frequently used to improve the collection efficiency of electrostatic precipitators by reducing the resistivity of the fly ash.[32B,111,112] The action of such agents is more complex than generally supposed, and various characteristics, such as mean particle size, may be affected.[113]

Ash Collection and Beneficiation

The characteristics of the fly ash depend upon the collector system and subsequent handling and storage procedures. Current practice has been reviewed with respect to the effect various operations have on the properties of the fly ashes.[92,114] Some problems and new developments have been described by Power.[115] Currently available particulate control systems include mechanical collectors, fabric filters, electrostatic precipitators (ESP's), and wet scrubbers. Efficiency of collection varies with particle size for each specific type of collector system, as shown in Table 2.3.[35B] Mechanical and electrostatic methods are most commonly used. Dry ESP's have overall efficiencies much higher than even high-efficiency cyclones. In addition, ESP's collect the coarsest particles first, and progressively finer particles are collected along the gas stream path.[31B] The particles are collected in several (typically three to five) separate hoppers which yield different particle size ranges and which may or may not be separately discharged for use.

If the ash is collected by one of the dry methods, it is transported by vacuum or pressure lines to a storage silo or by sluice lines to an ash pond. If the ash is collected by a wet collector, it will be sluiced directly to a pond for dewatering or disposal. The use of water will result in dissolution of some of the fly ash particles and the soluble deposits on the surfaces of other particles.

In some cases the fly ash collected at the power plant may be consistently able to meet existing specifications for use in concrete only if subjected to modest processing for beneficiation.[116,117] For example, in 1964 the Southern Fly Ash Company installed a Raymond separator at its plant in Wilsonville, Alabama. The separator recovered 84% of the fly ash and yielded a product averaging 92% passing the No. 325-mesh sieve.[23] More complex separation systems are being considered to produce several fractions for different uses.[118]

In the manufacture of blended cements with fly ash, the ash is sometimes interground in the finish mills with the portland cement.[119,120] Grinding not only reduces the maximum particle sizes but also increases the reactivity.

In this chapter we have seen how coal composition and firing can affect the composition, morphology, and size of fly ash particles. Before showing how these characteristics affect the performance of fly ash in concrete, the composition, particle morphology, and physical properties will be examined in detail in the next two chapters.

Table 2.3 Several Operating Characteristics of Particulate Collectors
(Table from Ref. 35B)

General class	Specific type	Typical capacity	Overall efficiency (%)	Fractional efficiency in percent for various size ranges in μm 0-5	5-10	10-20	20-44	>44
Mechanical collectors	Baffle	1000-3500 ft^3/min per ft^2 of inlet area	60	7.5	22	43	80	90
	Conventional cyclone		65	12	35	57	82	91
	High-efficiency cyclones	2500-3500 ft^3/min per ft^2 of inlet area	85	40	79	92	95	97
Fabric filters	Automatic	1-6 ft^3/min per ft^2 of fabric area	99+	99.5	100	100	100	100
Wet scrubbers	Impingement baffle	400-600 ft^3/min per ft^2 of baffle area	—	—	—	—	—	—
	Packed tower	500-700 ft^3/min per ft^2 of bed cross-sectional area	94	90	96	98	100	100
	Venturi	6000-30,000 ft^3/min per ft^2 of throat area	99+	99	99.5	100	100	100
Electrostatic precipitators	Dry, single-field	2-8 ft^3/min per ft^2 of electrode collection area	97	72	95	97	99+	100
	Wet (charged-drop scrubber)	5-15 ft^3/min per ft^2 of electrode collection area	—	—	—	—	—	—

Source: Adapted from K. Wark and C. F. Warner. *Air Pollution, Its Origin and Control,* New York: IEP/A Dun-Donnelley Publisher, 1976.

1 $ft^3/min/ft^2$ = 0.00508 m/s

Chapter 3

Fly Ash Composition and Particle Morphologies

General Description

Fly ashes are heterogeneous fine powders consisting mostly of rounded or spherical glassy particles of variable silica, alumina, and iron oxide content. There are also irregular or angular particles; these include both unburned coal remnants and mineral particles. Depending on the source, fly ashes may be finer or coarser than ordinary portland cements. ASTM C618 specifications[1] require that no more than 34% be retained on the No. 325 (45 μm) sieve. Figs. 3.1 through 3.3 are scanning electron microscope (SEM) micrographs centered at the same point of one sample of fly ash at three magnifications. The lowest magnification shows a large field of view which includes a representative number of the largest particles. The higher magnifications show small particles and some details of the structure of the surfaces.

Figs. 3.4 and 3.5 are optical micrographs of thin sections of another fly ash in an epoxy mount. Some of the spherical particles are seen to be solid. One relatively large particle in Fig. 3.4 appears to be a portion of a thin hollow sphere that contains many smaller particles. Fig. 3.5 shows a large irregular carbon particle and several spherical particles which appear to be polycrystalline.

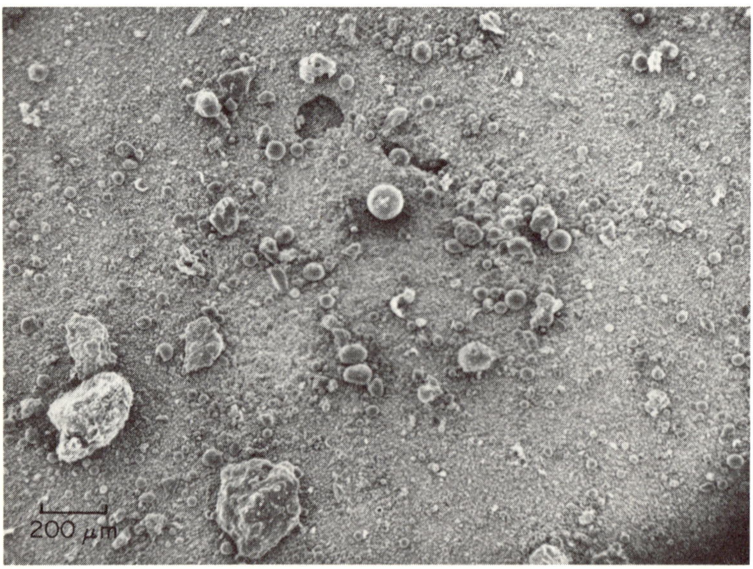

Fig. 3.1 Crawford Edison (Chicago) Class F fly ash at low magnification (65X). Bar is 200 μm. PCA micrograph by H. Love.

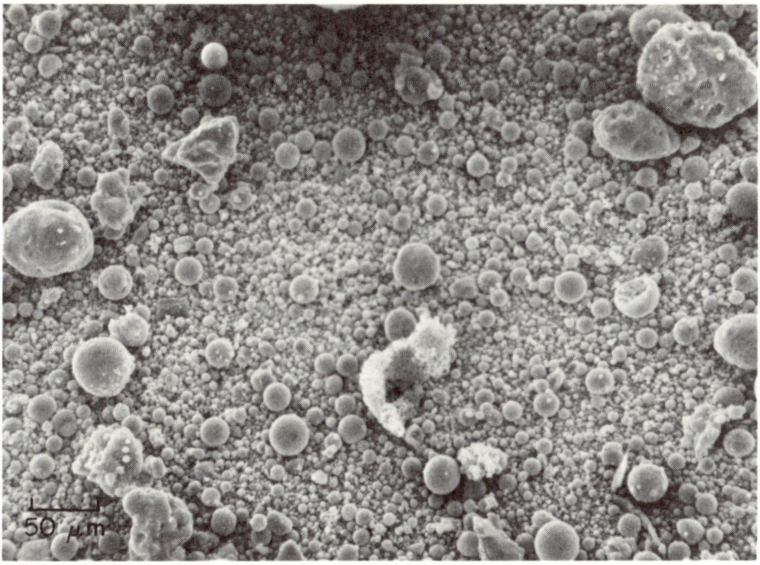

Fig. 3.2 Crawford Edison (Chicago) Class F fly ash at 260X. Bar is 50 μm. PCA micrograph by H. Love.

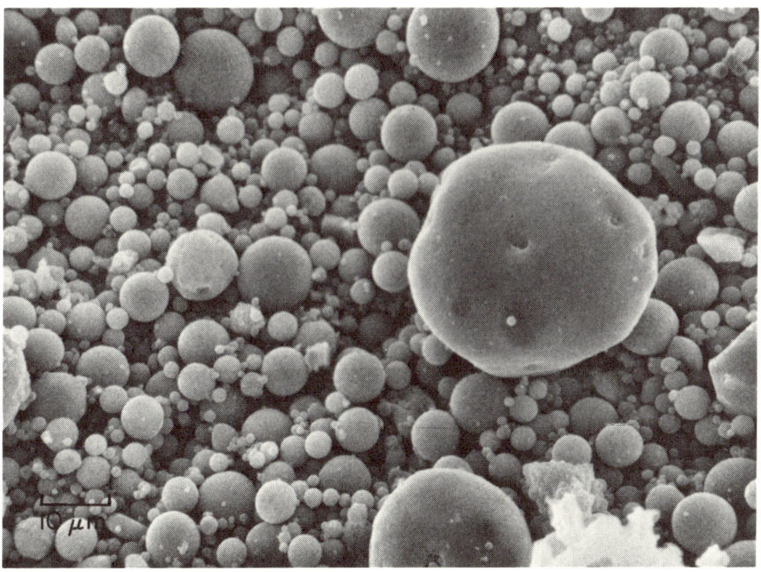

Fig. 3.3 Crawford Edison (Chicago) Class F fly ash at 1300X. Bar is 10 μm. PCA micrograph by H. Love.

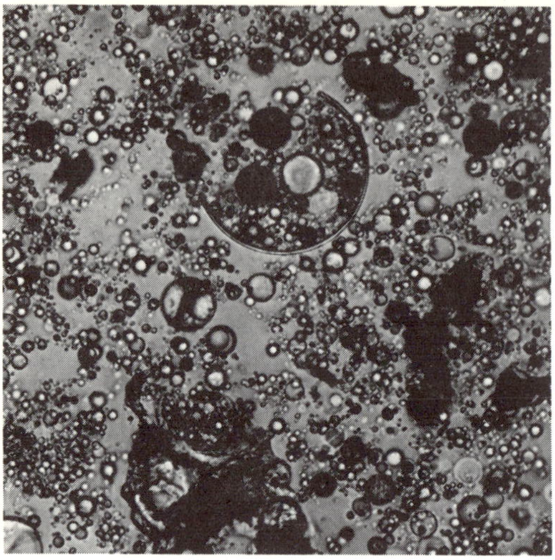

Fig. 3.4 Jeffries (St. Marys, Kansas) Class C fly ash at 200X (field is 210 μm). Optical micrograph of thin (about 20 μm thick) section of epoxy mount showing large thin-shelled broken sphere containing smaller particles (plerosphere). PCA micrograph by D. H. Campbell.

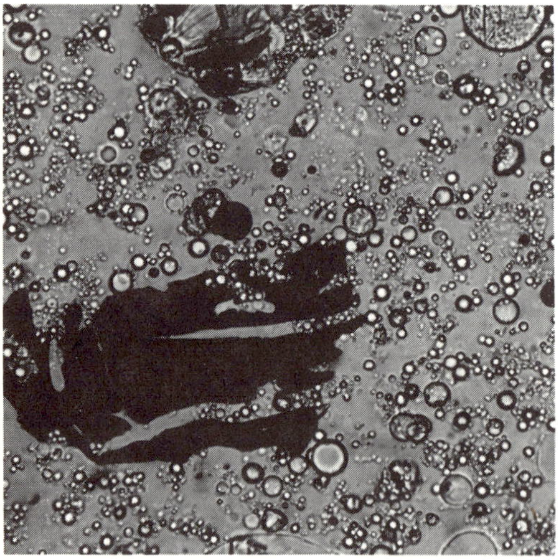

Fig. 3.5 Jeffries (St. Marys, Kansas) Class C fly ash at 200X (field is 210 μm). Optical micrograph of thin (about 20 μm thick) section of epoxy mount showing large carbon particle. PCA micrograph by D. H. Campbell.

Bulk Chemical Composition

Data given in Table 2.2 for coal ash compositions for different coals indicate what compositions might be expected for the fly ash produced by combustion of pulverized coals. However, there are several reasons for deviations from the whole coal ash composition. Fly ash may differ in composition from the bottom ash or slag, especially for minor and trace elements.[33B] The more volatile constituents are not very likely to be retained in the furnace, but they may condense on fly ash particles or escape up the stack. Also, the addition of conditioning agents to improve performance of the electrostatic precipitators may affect fly ash composition and other properties.

Wide variations in coal compositions, combustion conditions, and ash collection systems result in wide variations in fly ash compositions. Many reviewers have summarized ranges of compositions commonly encountered, and a few of these are given in Table 3.1 to illustrate the nature of the variations of major constituents. Berry[41B] included a substantial number of individual analyses in his review. Parker and Ray[33B] emphasized the trace element analyses in their review and included descriptions of many modern methods of analysis.

Dunstan[62a] compared the oxide compositions of nine fly ashes with those of blast furnace slags and portland cements, as shown in the CaO-SiO_2-Al_2O_3 ternary diagram in Fig. 3.6. From their high silica contents and relatively low CaO contents, the alumino-silica fly ashes in the mullite ($3Al_2O_3 \cdot 2SiO_2$) field may be expected to react with calcium hydroxide just as do natural pozzolans. Other fly ashes, with higher calcium contents, lie in the anorthite ($CaO \cdot Al_2O_3 \cdot 2SiO_2$) and gehlenite ($2CaO \cdot Al_2O_3 \cdot SiO_2$) fields closer to the composition of slags and portland cement. Such diagrams should not be taken too literally because of the complicating effects of other constituents in the fly ash, e.g., Fe_2O_3 and alkalies, which may exist in solid solution in the glass and also in separate

Table 3.1 Typical Compositions of Fly Ashes As Summarized by Several Reviewers, %

	Abdun-Nur[36B] 1961	Jarrige[31B] 1971		Sersale[127] 1980	
Component	Coal	Coal	Lignite	Bituminous coal	Lignite
SiO_2	28-51	47-54	18-25	50	23-50
Al_2O_3	15-34	28-35	12-15	30	8-14
Fe_2O_3	4-26	4-12	6-8	7	8-20
CaO	1-11	1-4	43-49	2	18-50
Free CaO	—	0.1	18-25	—	—
MgO	0-3+	1-2.5	2-3	5	—
Na_2O	—	0.2-2	5 (Na₂O + K₂O)	5	—
K_2O	—	1-6			—
SO_3	0.2-4	0-1	5-9	—	—
C	—	0.5-12	1-3	1-6	—
LOI	0.56-32	2	1	5-12	—

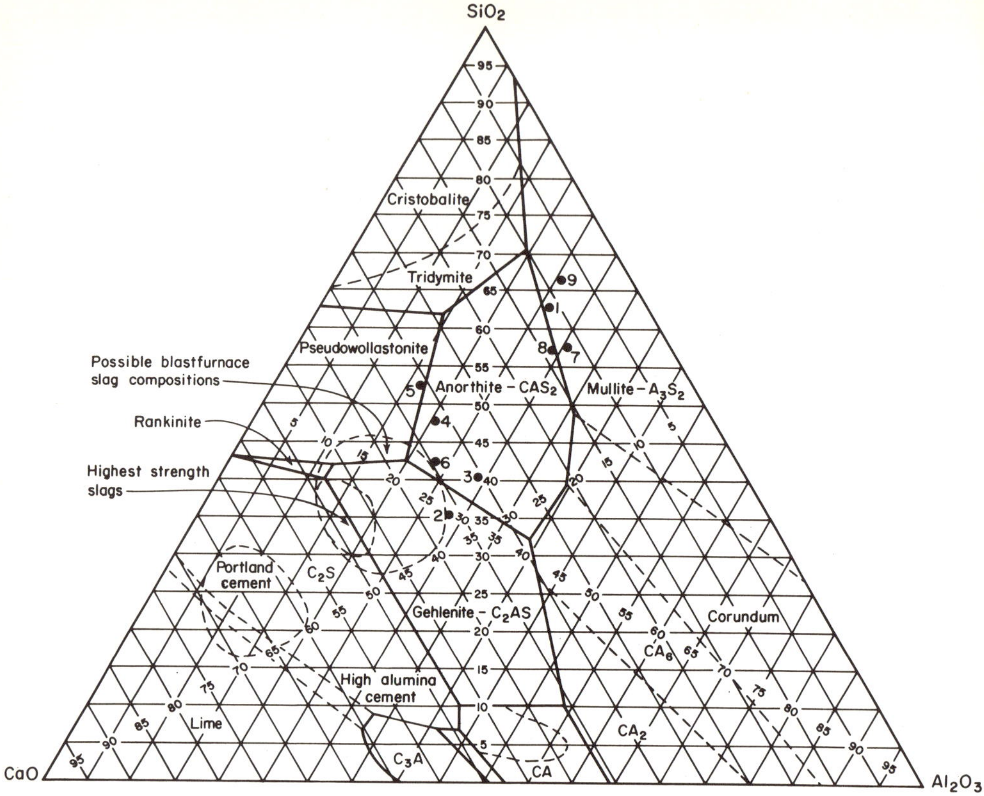

Fig. 3.6 Comparison of compositions of nine fly ashes (at numbered points) with blast furnace slags and portland cements in the system $CaO–SiO_2–Al_2O_3$. (Ref. 62a)

crystalline phases. Nevertheless, mullite is one of the major crystalline phases to form from the molten glass during cooling of the fly ash particles in the Class F range of compositions.

Phase Analyses

Light microscopic techniques and X-ray diffraction (XRD) have been used by many workers to identify and determine the quantities of crystalline and other phases present in fly ash.[121–126] Luke[121] reported the results of such mineralogical examinations to determine the amounts of various phases in five particle size fractions of two different fly ashes. One fly ash was collected with an electrostatic precipitator and was finer than the fly ash collected with a mechanical collector. Representative samples of the fly ashes were wet-sieved over No. 200, 325, 400, and 500 sieves. Each fraction was examined with the petrographic microscope to obtain the results in Table 3.2.

Table 3.2 Mineralogical Compositions of Particle Size Fractions of Two Fly Ashes[121]

	Electrostatically precipitated					
	% retained on sieves*					
Constituent	74 μm No. 200	44 μm No. 325	37 μm No. 400	25 μm No. 500	% passing No. 500	Whole sample, %**
Glass	32	49	52	56	87	79
Magnetite-hematite	2	14	13	14	5	6
Carbon	33	8	9	5	1	4
Anisotropic material	27	22	18	15	3	6
Aggregates	6	7	8	10	4	5
Totals	100	100	100	100	100	100

	Mechanically collected					
	% retained on sieves*					
Constituent	74 μm No. 200	44 μm No. 325	37 μm No. 400	25 μm No. 500	% passing No. 500	Whole sample, %**
Glass	35	55	61	58	63	58
Magnetite-hematite	8	5	20	26	16	16
Carbon	47	29	7	8	5	13
Anisotropic material	3	3	3	3	5	4
Aggregates	7	8	9	5	11	9
Totals	100	100	100	100	100	100

*Percentage is based on count of more than 300 particles in each sieve fraction.
**Percentage is based on gradation of as-received sample and on distribution of constituents of wet sieved fractions.

The most abundant phase in each fly ash and on each sieve was glass. Birefringent material in the glass was not identified but was probably mullite, since mullite was the only crystalline phase other than the hematite, magnetite, quartz, and calcite (calcite was found to be a contaminant) found in the XRD patterns of the same material. Magnetite was the second most abundant constituent in both fly ashes. Very little hematite could be identified microscopically, but it was found in relatively large amounts in the X-ray patterns; therefore, it was concluded that the magnetite particles identified under the microscope were actually mixtures with hematite.

The anisotropic material included the low birefringent material enclosed in glass particles (probably mullite and partially devitrified glass) and well-crystallized quartz particles. Agglomerated glass, magnetite, and carbon particles were found in each size range. Most of these were easily broken up.

Data in Table 3.2 show that in these fly ashes there was a definite phase separation according to particle size for the glass and carbon particles. The finest fractions were richest in glass. Carbon was most abundant in the coarsest fractions.

In an investigation of 14 fly ashes collected with electrostatic precipitators, Watt and Thorne[(123)] found by microscopic and XRD examinations that most fly ashes, after extraction with water, contained only four crystalline phases in significant amounts: quartz, mullite, hematite, and magnetite. This finding is in agreement with the results of other workers for aluminosilica (Class F) fly ashes. Seven of the fly ashes were analyzed for these minerals by X-ray methods. From these analyses and the chemically determined carbon contents, the glass contents were determined by difference, yielding the results in Table 3.3. Then, from the oxide chemical analyses, the average compositions of the glass phases were calculated. These are given in Table 3.4.

Table 3.3 shows again that the glass is by far the most abundant constituent in all of these fly ashes. Amounts of the mineral phases are rather small and do not appear to be directly related to the pozzolanic properties of these ashes, except as they influence the amount and composition of the glass phase. The glass phase compositions in Table 3.4 differ little, except for the high CaO content of the Rye House ash. Lime contents of fly ash will be discussed later in terms of the existence of another (cementitious) phase. None of the four major crystalline phases is either hydraulic or pozzolanic.

Table 3.3 Mineralogical Analyses of Fly Ashes,[(123)] wt %

Fly Ash	Quartz	Mullite	Hematite	Magnetite	Carbon*	Glass**
Dunston (1)	4.5	11	2.7	1.4	3.1	77
Dunston (2)	3.8	11	1.9	1.2	2.0	80
Ferrybridge (1)	2.8	6.5	1.6	1.9	1.5	86
Ferrybridge (2)	2.2	6.5	1.1	0.8	0.9	88
Hams Hall	3.5	10	1.6	1.2	0.6	83
Rye House	8.5	14	2.1	2.5	2.4	71
Skelton Grange	4.1	10	2.4	2.6	2.1	79

*Determined chemically
**Determined by difference

Table 3.4 Calculated Mean Compositions of Glassy Components of Fly Ashes,[(123)] wt %

Fly Ash	SiO_2	Al_2O_3	Fe_2O_3	CaO	MgO	Na_2O	K_2O	TiO_2
Dunston (1)	56	29	5.9	1.3	2.6	0.3	3.5	1.5
Dunston (2)	55	33	4.2	1.4	2.1	0.3	2.1	1.5
Ferrybridge (1)	53	28	7.3	1.6	1.9	1.6	5.0	1.1
Ferrybridge (2)	52	27	10	0.8	2.4	1.8	4.6	1.1
Hams Hall	55	27	6.4	3.2	2.7	1.0	3.8	1.3
Rye House	48	22	11	9.7	4.4	0.8	3.2	1.5
Skelton Grange	53	26	9.1	3.2	2.5	1.0	4.8	1.0

As noted earlier, high calcium fly ashes may contain appreciable free CaO. Some of the CaO may be encapsulated in glass.[128] Instead of mullite, fly ashes may contain minor amounts of other crystalline phases characteristic of their compositions.[62,124–129] Melilite (a Na-Ca-Mg-Al silicate), gehlenite (Fig. 3.6), kalsilite, the anhydrous calcium sulfoaluminate ($3CaO \cdot 3Al_2O_3 \cdot CaSO_4$), calcium sulfate, alkali sulfates, and the major compounds in portland cement have all been detected by XRD in such fly ashes.

XRD also reveals a significant difference between the broad "hump" in the patterns for the glassy phases in the high-calcium and low-calcium fly ashes.[125,130] The low-calcium glasses yield a maximum intensity in the pattern at about 24° 2θ (Cu radiation), which is characteristic of most siliceous glasses. The high-calcium fly ashes yield a highly skewed pattern that peaks at about 32° 2θ. For CaO contents from 0 to 20%, the position of the maximum was found to vary with the % CaO according to

$$2\theta = 22.7 + 0.24(\%\text{CaO}) \tag{3.1}$$

At higher CaO contents the position shifted abruptly from about 27° to 34° 2θ, where it remained constant up to about 26% CaO.

Particle Morphologies

Many investigations have shown that fly ashes are extremely heterogeneous both physically and chemically. However, Watt and Thorne[123] found that most of the particles could be classified into just a few major types. The range of variation of composition and physical properties within each type was relatively small. Lesser amounts of a few other types of particles, and the tendency to form agglomerates, made the fly ashes appear to be more heterogeneous than indicated by subsequent analysis.

Watt and Thorne supposed from what was known of the distribution of minerals in pulverized coal, and of the conditions of combustion in the furnace, that each primary fly ash particle is probably produced by the burning out of a single coal particle containing only one, and almost certainly not more than two, mineral species. (Cf. Malte and Rees in Chapter 2, "Combustion and Ash Formation.") They then confirmed this supposition by producing synthetic fly ash particles in a flame from 12 pure minerals representative of the main minerals found in coal. These included several clay minerals, carbonate (dolomite), sulfide (pyrite), and accessory silicate minerals, including quartz and feldspar. These synthetic fly ash particles closely resembled those found in the commercial ashes.

The 14 commercial ashes examined by Watt and Thorne were each chemically analyzed as a bulk material, examined microscopically, and then subjected to several separation processes to disperse agglomerates and to classify the ashes by particle size, magnetic properties, and density.

In most cases it was found that the agglomerates could be broken down into their constituent primary particles by extracting the ash exhaustively with water. In one ash the agglomerated particles appeared to be sintered and were not broken down by water extraction. The water-soluble content ranged from 1.7 to 4.7% of the ash. Microscopic examination showed that the ashes were much less heterogeneous after extraction than before.

The particles could then be readily classified into the eight different types given in Table 3.5. Most of the fly ash in all 14 fly ashes was comprised of particles of Types 1, 2, and 3. Of these three major types, the Type 2 particles were the least abundant and most heterogeneous; most of them could be removed by magnetic separation. Each ash was made up mostly of Types 1(a) and 3, clear glass spheres and rounded spongy glassy particles, respectively, although the proportions varied with different fly ashes. The clear glass spheres were generally smaller than other particles; they are characteristic of ashes with high specific surface areas. Some of the Type 1 particles contained crystalline matter (mullite) ranging from one or two minute needlelike crystals to nearly complete crystallization. The Type 3 particles had both small and large bubbles and ranged from foam to cenospheres. The spongy particles were larger and were generally the most characteristic feature of the ashes of low specific surface area. Crystallinity of these particles was very limited.

The particles of Type 4 were present in sufficient quantity to have a significant effect on the properties of only one of these fly ashes. These particles were easily recognizable but showed no clear evidence of their origin. Tests showed that they were highly reactive and that their average CaO content was about 40%. Their average composition was comparable with that of a hydraulic cement.

The Type 5 particles were apparently sintered agglomerates which generally contained red (Type 8) hematite particles. The largest particles (up to 500 μm) were of this type.

The Type 6 particles were residual, partly-burned fragments of coal. A large amount of this material appeared to be associated with a high content of the spongy particles, suggesting that the spongy particles are a result of low furnace temperatures or short residence times and incomplete combustion.

The angular Type 7 particles (probably quartz) were presumably derived from quartz originally present in the coal.

In the 14 fly ashes studied by Watt and Thorne, 4 to 16% of the ash could be separated out magnetically. These particles were opaque spheres and rounded Type 2 particles shown to be composed of magnetite and hematite. About half of the total iron in the ash was in these particles.

Fisher et al.[131] made a detailed study of four size fractions of fly ash collected at a Western U.S. power plant burning low sulfur, high moisture, and high ash content coal. A classification system based on 11 morphological classes was developed in order to determine the relative abundance of the particle types in the different size fractions. Although the samples were collected downstream of the electrostatic precipitator system, they contained essentially the same particle types described by Watt and Thorne. Figure 3.7 illustrates these particle types and their probable genesis. Volume median diameters of the four size fractions were 20, 6.3, 3.2, and 2.2 μm, as determined by Coulter counter. The relative abundance of the 11 morphological particle types in the coarsest fraction was in general agreement with that reported by Watt and Thorne. The samples also had

Table 3.5 Types of Fly Ash Particles Characterized by Microscopic Examination[123]

Type	Shape	Color	Crystallinity and texture	Characteristic size range, μm	Comments
1	Spherical and rounded	Colorless	a. Glassy, clear, solid b. Glassy, containing small bubbles c. Glassy, with crystal traces d. Predominantly crystalline, solid	0-20 10-50	
2	Spherical and rounded	Light brown to black	Lighter-colored ones, glassy, all solid	5-30	Deepening color suggests increasing iron content
3	Rounded	White in reflected light	Glassy, spongy	10-200	Small and large bubbles giving a range from foam to cenospheres
4	Irregular	Light brown	Partly crystalline, solid	10-100	Irregularity of profile and surface very marked
5	Irregular	Varicolored in reflected light	Party crystalline, solid	50-500	Agglomerated particles, apparently sintered; generally containing red particles and areas
6	Irregular	Black	Solid or porous	20-200	Partially burned coal particles, sometimes containing mineral matter
7	Angular	Colorless	Crystalline, solid	10-100	Probably quartz
8	Angular	Red	Crystalline, solid	5-50	Probably hematite

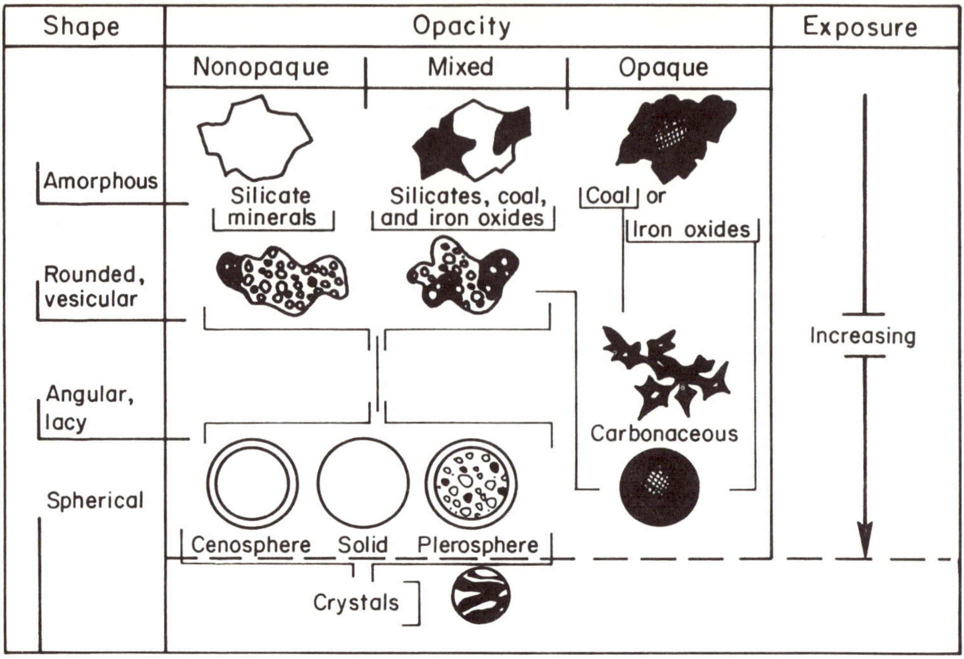

Fig. 3.7 Fly ash morphogenesis scheme illustrating probable relationship of opacity to particle composition and relationship of particle shape to exposure in combustion chamber. (Ref. 131)

a very high concentration (87 vol. %) of the solid nonopaque spheres in the finest fraction.

Light microscopic and SEM examinations showed that, in addition to the presence of "quench" crystals in the glassy spheres, there were acicular, elongated blades, and cuboid crystals of submicron size on fine fly ash particles. Some of the larger acicular crystals were found by electron microprobe analysis to be calcium sulfate.

Fig. 3.7 indicates the existence of cenospheres packed with smaller spheres. These packed spheres have been called plerospheres and are described in detail by Fisher et al.[132] as being more typical of fly ash from Western U.S. coals. Carpenter et al.[133] also examined samples of fly ashes obtained from burning both Midwestern and Western U.S. coals with the express purpose of verifying the presence of plerospheres (misspelled pleurospheres) by crushing undamaged particles under the microscope. In no cases were filled spheres observed before crushing; it was suggested that when filled spheres are found, the smaller material may have filled the larger spheres after they were cracked or punctured during collection and transportation.

Detailed Phase and Surface Analyses

The composition of surface coatings on the fly ash particles, formed during cooling by condensation of volatilized constituents, has recently been the subject of considerable research because of its environmental implications.

Four basic methods have been used to study the distribution of elements and compounds within or among fly ash particles:[109]

1. The total concentration of an element can be determined as a function of the particle size; if the concentration increases with decreasing particle size, it is often assumed that the element is concentrated on the particle surfaces.
2. The surfaces of the particles can be analyzed directly by the use of modern spectroscopic methods such as ESCA (electron spectrometry for chemical analysis, also called X-ray photoemission spectroscopy), AES (Auger electron spectrometry), and SIMS (secondary ion mass spectrometry).
3. Extraction or leaching with various solvents distinguishes the compositions of soluble from insoluble phases if new phases are not formed by precipitation.
4. Single particles may be analyzed by secondary X-ray emission in an electron microscope.

Hansen and Fisher[109] examined the distribution of major, minor, and trace elements between the aluminosilicate matrix phases and surface phases of the same four particle size classified fly ashes referred to earlier.[131] These fly ashes were collected in a Western U.S. power plant burning low sulfur, high moisture, high ash content coal. Two techniques were used to determine the concentrations of elements in the matrix and the surface material, namely, the dependence of total element concentration on particle size and the solubility of elements in acid solutions of various concentrations. Although most of the work concerned trace elements, the methods appear to be useful for more abundant elements as well.

For particle size dependence, the total concentration C_T of each element is expressed as

$$C_T = C_{\mathrm{M}} + C_{\mathrm{A}} \frac{6}{D_V \rho} \tag{3.2}$$

where C_{M} is the concentration in the matrix, C_{A} is the surface concentration (per unit area), and D_V is the volume-mean diameter. (D_V is the diameter of a sphere that would have the same specific surface as the actual size fraction.) If C_{M} and C_{A} are constants, a plot of C_T versus $6/D_V\rho$ should be linear. The intercept at infinite diameter then yields C_{M}, and the C_{A} can be obtained from the slope.

When the experimentally determined C_T values were plotted versus $6/D_V\rho$, it was found for some elements (e.g., Al, Na, K) that there was no significant correlation with particle size. For some other elements, there was a good linear relationship that showed zero concentration at infinite particle size; presumably, these elements (As, Cd, Sb, and W) were almost entirely concentrated in surface phases. Several elements, including Ca and Si, showed a slight to intermediate dependence on particle size. Although these results do not seem entirely consistent with other data discussed later, the method does seem to yield consistent results for some elements.

There were several difficulties with the two acid leaching (HCl and HF) methods which required that various corrections be made; for some elements, discrepancies between the three methods remained. It was possible, however, to conclude that more than 70% of the Na, K, Mg, and Fe was associated with the aluminosilicate matrix. More than 70% of the Ca appeared to be associated with the same particle size distribution as the aluminosilicates. These phases may be free CaO and the cementitious compound of Watt and Thorne's Type 4 particles.

Linton et al.[107] studied the surface composition of a fly ash produced from a power plant burning pulverized Midwestern U.S. bituminous coal. Large particles (45 to 180 μm) were analyzed by SIMS and AES as practiced for microanalysis, i.e., ion microprobe mass spectrometry and scanning Auger microscopy. These experimental techniques, which are described in some detail, were used to determine the concentration versus depth profiles of various elements in the surface regions relative to the concentration at greater depths (0.1 μm). In addition, the surface concentrations after leaching in water were measured to determine the leachability of the surface constituents.

Certain elements (Si, Al, Ti, Ca, and Mg) were found not to be concentrated on particle surfaces; these elements also had relatively low leachabilities in water. However, K, Na, Fe, and S were found to be concentrated on particle surfaces and to be highly leachable. Surprisingly, the peak surface concentrations did not diminish significantly after leaching, an anomaly that was not fully explained. Dissolution of soluble phases may have prevented leaching from the surface. For unleached particles, Auger depth profiles showed qualitative agreement with the ion microprobe results in that only K, Na, and S exhibited major surface enhancement to a depth of about 0.003 μm; by ion microprobe techniques the surface enhancement was indicated to about twice that depth.

The application of high spatial resolution (3 μm) AES to the surface characterization of fly ash was investigated in some detail by Powell and Spicer.[134] The use of ion sputtering to remove surface material enabled the AES technique to be extended to greater depths (0.04 μm) than without the ion etching. Six fly ash samples of diverse bulk chemical compositions were examined. SO_3 contents ranged from 0.4 to 2.29%, the latter having 1.56% as soluble sulfate by ordinary methods. The Auger depth profiles showed the sulfur and also carbon to be concentrated in a surface region about 0.01 to 0.03 μm thick for each of the fly ash samples studied. One ash also had detectable amounts of chlorine and phosphorus on the particle surfaces. Surface concentrations of sulfur in two of the fly ashes were estimated to be about 7 and 10 atom %. Both of these values were far in excess of the bulk sulfur concentrations of 0.5 and 1.86%, respectively. Although the bulk concentrations were very different, the surface concentrations were more nearly the same.

These results were qualitatively confirmed by two other techniques, SIMS and ESCA. The ESCA spectrum identified the oxidation state of the sulfur as the sulfate form. The SIMS spectra before ion etching showed strong peaks for K,

Na, Al, Ca, Si, and Mg, but only a small peak for sulfur. Powell and Spicer pointed out that the relative sensitivity of SIMS to elements such as Na, Mg, and Ca is very large and that the strong peaks do not necessarily mean that the surface is composed of such compounds. The relative sensitivity of SIMS to sulfur is much smaller, so that without other techniques its high surface concentration may have escaped attention. This report appears to be a very useful guide for the characterization of fly ash by Auger spectroscopy and related modern surface analytical techniques. The relative merits of these methods are also discussed in some detail.

Phase compositions of four different fly ashes were examined in detail by Hulett et al.[135,136] using modern analytical methods (SEM-EDX and XRD) in combination with HF leaching of particles of selected types. The ashes were particle size classified after being washed with water to eliminate agglomerates. Most of the work was done with the 100-to 200-μm size fractions, but additional work with the less than 37-μm fraction indicated that the conclusions reached are true for most particle sizes. There were four main particle types: (1) spheres and spherical shards (transparent to opaque), (2) opaque oblong shapes like miniature pebbles, (3) quartz sand (nonfused), and (4) carbon cinders.

The Type 1 and Type 2 particles were etched by the 1% HF solution so that all of the aluminosilica glass phase was removed, leaving only crystalline skeletons in the original particle shapes. Only the white/opaque Type 1 and 2 particles remained after etching; none of the translucent/transparent spheres and shards remained. The crystalline skeletons were either mullite in acicular and chunky forms, quartz, or mixtures of these, as shown in Fig. 3.8. The acicular mullite had a composition approximating $3Al_2O_3{\cdot}2SiO_2$, with some Fe and Ti isomorphically substituted for Al and Si, respectively; Na, Mg, K, and Ca were completely removed with the glass phases during etching.

In addition to the original quartz sand (Type 3) particles, which were never in the molten state, the etched Type 1 and 2 particles revealed recrystallized silica that had a corallike morphology; XRD showed this to be mostly quartz.

The weight ratio of glass to crystalline phases in the 100 to 200-μm size glassy particles varied from 1.49 to 2.49 for the different ashes. This range corresponds to about 59 to 71% glass. One sample that contained 64% glass in this size fraction contained 71% glass in the less than 37-μm size fraction. For this finer fraction, the composition of the glass phase was calculated to be 34% Si, 5.86% Al, 4.3% Fe, 3.7% K, 1.7% Ti, 1.5% Ca, 1% Mg, 48% O; the Si to Al ratio in the glass was about 5.5.

Hulett and Weinberger[135] suggested that their findings provide an alternative to the common assumption that surface enrichment of certain elements occurs because of condensation from the vapor phase on the particle surfaces. Conceding that the condensation model may well be true for elements such as sulfur, they suggest that surface enrichment can also occur by diffusion from inside the particle during crystallization. It appears that such compositional differences and gradients resulting from crystallization could affect the pozzolanic activity of the glass phases. It has been reported that the most reactive particles are not entirely vitreous, but contain inclusions of mullite and quartz.[137]

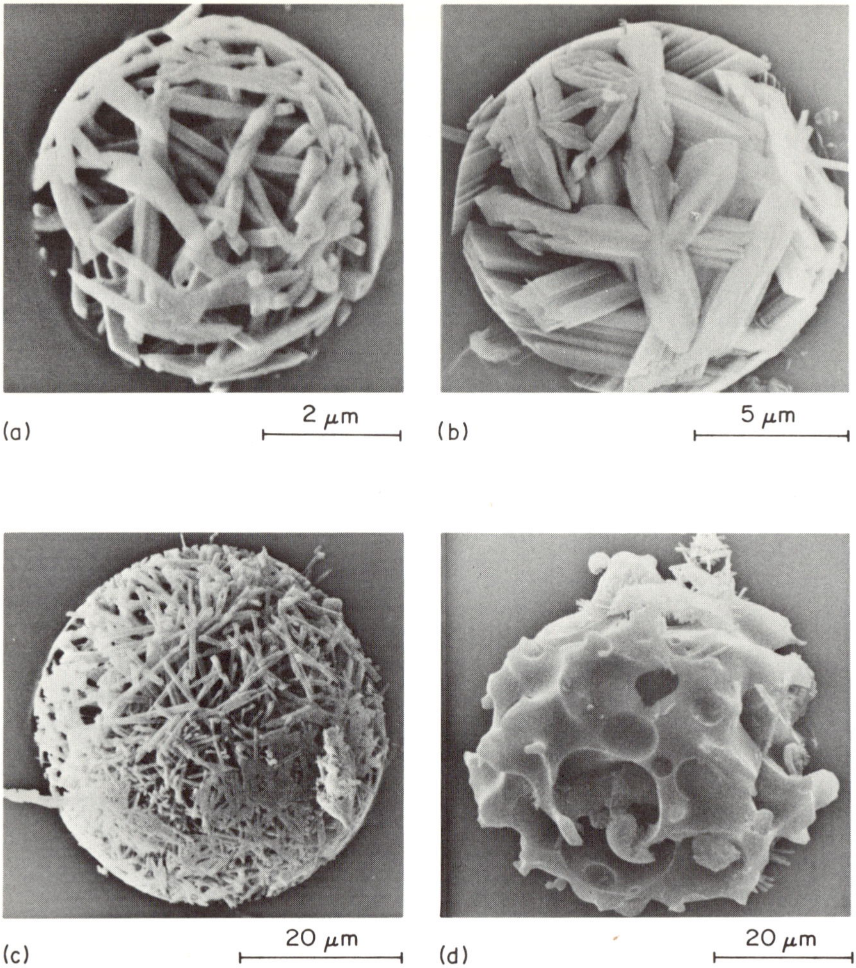

Fig. 3.8 Mullite and quartz skeletons of fly ash particles after etching with 1% HF solutions. (Ref. 135)

Adsorbed Organic Residues

In "Composition of Coal" in Chapter 2, it was pointed out that the complex coal molecules are assemblages of polycyclic aromatic and polynuclear aromatic carbon compounds. Air pollution control studies have led to the development of methods for the determination of the major polynuclear aromatic hydrocarbons (PNA's) and polycyclic aromatic hydrocarbons (PAH's), which can be obtained from extracts of particulate emissions, including fly ash.[138] Research on coal combustion has shown that polycyclic organic molecules are adsorbed onto fly ash particles as the temperature falls.[139] Decomposition of the same PAH's is decreased substantially by adsorption onto the fly ash particles. Other compounds, such as 9, 10-dimethyl-anthracene and 9, 10-dihydro-anthracene, undergo rapid degradation when adsorbed onto fly ash surfaces, the extent depending on the nature of the fly ash.[110]

It should be noted that there are considerable compositional and structural similarities between some of these compounds and some commonly used water-reducing admixtures. The salts of naphthalene formaldehyde sulfonic acids used in some "superplasticizers" (e.g., Mighty-150) are produced by sulfonation of naphthalene.[140] It appears possible that some of the organic compounds adsorbed on the fly ash may be effective as water reducing or air-entraining admixtures, or both. It is also known that unburned carbon particles in the fly ash are sometimes effective adsorbents of admixtures.[3] These characteristics, along with particle size and morphology, may have important effects on the water requirement for flow of blended cement, mortar, and concrete.

In the next chapter physical properties of fly ashes will be further examined before dealing directly with the flow properties in Chapter 5.

Chapter 4
Physical Properties of Fly Ashes

Densities of Particles

The presence of bubbles in the glassy particles and voids in the agglomerated particles causes wide variations in the particle densities. Compositional fluctuations, especially of iron and carbon contents, also cause differences in density. These variations have been studied in some detail and correlated with chemical composition and other properties.[31B,121-124,141] Fig. 4.1 shows weight percent versus density data for different fractions of a fly ash collected at five points before or in an electrostatic precipitator (ESP) collection system. The two coarser fractions, with mass median diameters of about 100 μm or more, had maximum amounts in the density range 1.5 to 2.0 g/cm^3 (Mg/m^3). The finer fractions, with mass median diameters of 50 μm or less, had maximum amounts in the density range 2.0 to 2.5 g/cm^3.

Watt and Thorne[123] obtained detailed weight versus density distributions of six of the 14 fly ashes studied. Heavy liquid separations were made up to 2.88 g/cm^3. Fig. 4.2 shows results for three ashes which had pronounced differences in amounts of material in the 2.5 to 2.6 g/cm^3 density range. Microscopic examination showed most of these particles to be clear glass spheres; chemical analyses of this density fraction from the different fly ashes showed remarkably little variation (less than 4% for SiO_2). Fractions of equal densities from different ashes were generally found to contain the same kinds of particles.

Minnick et al.[141] performed statistical analyses of data reported in five major experimental studies. In each set of data, the fly ash characteristic which yielded the highest correlation coefficient with specific gravity was the Fe_2O_3 content. Multiple regression analyses using the Fe_2O_3 content as the first independent variable showed that the No. 325 sieve residue (325S) and CaO contents were also significant variables. The combined data from the five studies were used to obtain Eq. 4.1 for the specific gravity in terms of the percentage Fe_2O_3 and 325S:

$$\text{Specific gravity} = 2.23 + 0.016(Fe_2O_3) - 0.006(325S) \qquad (4.1)$$

The standard error of the estimate was given as 0.06.

The average density of the solid phases in fly ash can be measured after grinding the ash to high fineness (approximately 1,200 m^2/kg) to eliminate the bubbles and voids in agglomerates. Such values then range from 2.65 to 2.80 g/cm^3.[31B]

The average density of the solids without grinding varies between about 1.90 and 2.80 g/cm^3. The bulk density varies greatly with packing; without close compaction, values of 0.54 to 0.86 g/cm^3 are reported.[41B] Hobbs used values of 1.12 to 1.49 g/cm^3 for close-packed bulk densities.[142]

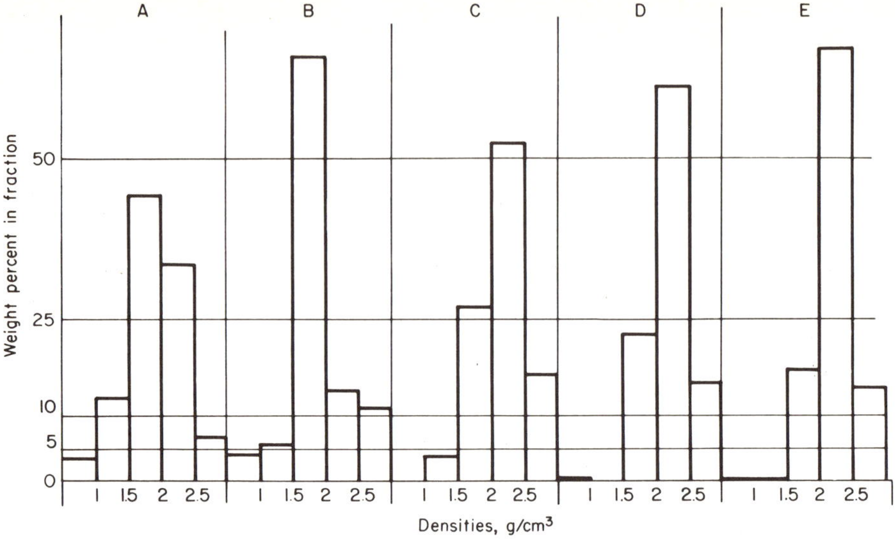

Fig. 4.1 Density distributions of fly ash samples of different size fractions collected at different points before or in an electrostatic precipitator collection system. (Ref. 31B) Approximate size ranges (and mass median diameters) are

A. 1000-20 μm (120 μm)
B. 800-15 μm (100 μm)
C. 600-15 μm (50 μm) ESP1
D. 300-1 μm (30 μm) ESP2
E. 200-1 μm (25 μm) ESP3

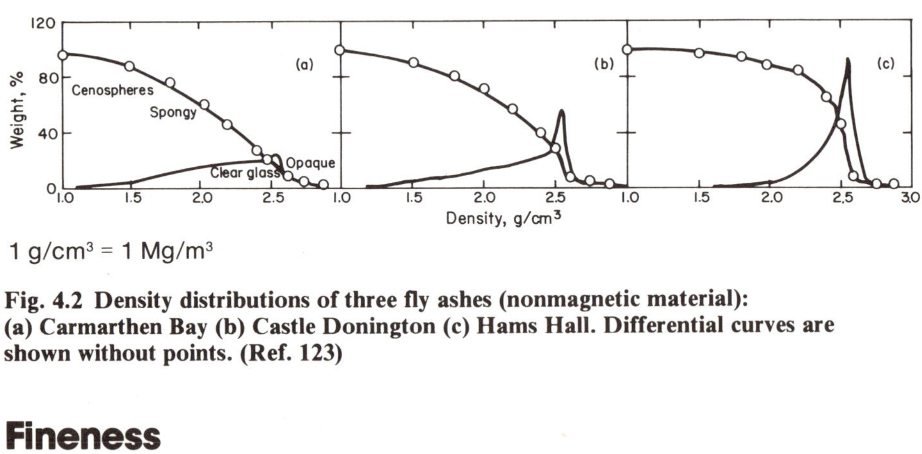

Fig. 4.2 Density distributions of three fly ashes (nonmagnetic material): (a) Carmarthen Bay (b) Castle Donington (c) Hams Hall. Differential curves are shown without points. (Ref. 123)

Fineness

The importance of the fineness of natural pozzolans and fly ashes with respect to their use in blended cements and concretes has long been recognized. Some measure of fineness is included in virtually all standard specifications.[3,76-78] Unfortunately, the fineness of fly ash has proven to be one of the most difficult properties to measure reliably and thus to specify in a meaningful way. Moreover, even fly ashes from a single source can be extremely variable over both short and long time periods. Therefore, it is essential to provide reliable methods to make meaningful measurements of the fineness of fly ashes.

One of the simplest and most reliable measures of fineness is the amount either retained on or passing through a certain size sieve. ASTM C618 currently specifies that for Class F and C fly ashes not more than 34% be retained on a No. 325 (45-μm) sieve. This single number does not, however, tell us very much about the material. More information is given by a cumulative particle size distribution curve which shows the amount finer (or coarser) than any particular size over the entire range of sizes.

Specific surface area is another very useful measure of fineness. Very fine particles have high specific surface areas. For example, a powder consisting of 1-μm diameter spherical particles of the density of portland cement would have a specific surface area of about 2,000 m^2/kg, about six times that of ordinary portland cements. For materials with a wide range of particle sizes, most of the specific surface area is contributed by the finest fractions.

Until 1977 a specification for the specific surface of fly ash was included in ASTM C618, but this was later deleted. There is considerable lack of agreement between particle size distributions and specific surface areas of fly ashes obtained by different methods. This discrepancy seems to be much truer for fly ash than it is for portland cement. It has been shown that surface area measurements are influenced by both the external and internal surfaces of the carbon particles.[143-145] Even for portland cements, specific surface areas by the Wagner Turbidimeter method (ASTM C115) do not agree with the value obtained by the Blaine air permeability method (ASTM C204). Not only do the properties of fly ashes cause measurement problems, but there are also inadequacies in some of our standard methods of measurement.

Discrepancies between specific surface areas measured by four different methods are shown in Table 4.1 for nine different fly ashes.[145] The air permeability values given are generally several times those calculated from the particle size distributions. However, these same air permeability values are seen to be but a

Table 4.1 Specific Surfaces of Fly Ashes Determined from Air Permeability, Nitrogen Adsorption, and Particle Size Distribution[145]

	Specific surface (m^2/kg)			
Ash code	Calculated from particle size	Air permeability	N_{dry}	N_{burned}
A	81	305	4070	760
B	97	413	3820	870
C	115	336	1020	570
D	92	209	480	1000
E	nd*	193	4700	1140
F	102	671	8900	670
G	81	311	6500	800
H	nd*	288	1240	680
I	80	254	970	660

*Not determined

fraction of most of those determined by nitrogen adsorption, even after ignition to remove unburned carbon. In one unexplained case (fly ash D), ignition increased the nitrogen surface area. The fact that the nitrogen adsorption areas were higher was attributed to the lack of measurement of internal fissures and pores in the particles by the air permeability method. However, the results showed that the specific surface areas by both air permeability and nitrogen adsorption on the dried samples correlated positively with the amount of unburned carbon.

Somewhat similar discrepancies were reported by Luke,[121] who determined both specific surface areas and particle size distributions for the same two fly ashes used to obtain the data of Table 3.2. Specific surface data obtained by air permeability, turbidimeter, and hydrometer (ASTM D422) methods are given in Table 4.2. As with the data in Table 4.1, these methods also yielded wide discrepancies between the values obtained on each fly ash.

A few years later Chopra and Narain[144] made a detailed investigation of specific surface area measurements of fly ashes. They concluded that the presence of +325 mesh material and unburned coal caused large discrepancies in such determinations. They also developed a modified procedure to obtain improved agreement between air permeability and turbidimeter values. The modified procedure included wet sieving (No. 325) and ignition of the fine fraction at 700° C; the ignited fine fraction was used for air permeability and turbidimeter measurements, which were summed with the specific surface area obtained for the coarse fraction calculated from its sieve analysis. When this modified procedure was used, the ratio of the air permeability to the turbidimeter specific surface areas in all cases was about 1.8, the same as that usually found for portland cements when turbidimeter surface area is calculated in the usual way.

Usually the turbidimeter specific surface area is calculated by taking the weight fraction of particles less than 7.5μm to be 3.75 μm in diameter. Chopra and Narain also made determinations of the weight fractions down to less than 1.5 μm and calculated the specific surface area by taking the fraction less than 1.5 μm to be 0.75 μm in diameter. These values were about twice as high as those calculated in the usual way and are much closer to the values obtained by the air

Table 4.2 Specific Surface Areas of Two Fly Ashes Determined by Air Permeability, Turbidimeter, and Hydrometer [121]

Method	Mechanical Collector	Electrostatic Precipitator
Air Permeability	116.7 m^2/kg	348.0 m^2/kg
Turbidimeter	72.0 m^2/kg	157.0 m^2/kg
Hydrometer	107.9 m^2/kg	312.6 m^2/kg
Density	2.41 g/cm^3	2.52 g/cm^3

1 g/cm^3 = 1 Mg/m^3

permeability methods. These results confirm other indications that the usual turbidimeter calculation systematically underestimates the contribution of the finest particles to the specific surface.

It should be noted that sieving and water extraction during wet-sieving are both effective means of dispersing agglomerates. The effect of agglomerates on the values of specific surface areas of finely ground pozzolans was investigated in detail by Alexander.(146) In addition to Alexander, others(144) have found it necessary to modify the porosity function used in the air permeability method in order to obtain consistent values of specific surface areas. Such modifications are especially important for high fineness fly ashes, because such materials cannot be compacted to the porosity normally used (0.5) with the air permeability method. In 1984 ASTM adopted a modified porosity function in C 204 and a special procedure for use with materials other than portland cements, such as fly ash.(21B)

Some of the particle size distributions determined by Luke(121) are shown in Figs. 4.3 and 4.4 to illustrate discrepancies between different methods for the same two fly ashes of Tables 3.2 and 4.2.

Wet-sieve analysis data are given from 500 μm down to 25 μm and are thought to be least subject to error. The Wagner turbidimeter, Klein hydrometer, and microscopic count methods all approximately agree with these sieve data down to about 30 μm, but diverge for smaller particle sizes. The turbidimeter and hydrometer curves appear to show fines contents much too low to account for the air permeability values; the microscopic count values are much higher, especially for the finer fly ash. If the microscopic count results are correct for the finer ash, it appears that the specific surface would have to be much higher than the air permeability value of 348 m^2/kg given in Table 4.2 (Cf. Fig. 4.6).

Some of these discrepancies seem to have been greatly reduced by the procedures used by Watt and Thorne.(123) Particle size distributions of the ashes that were disagglomerated by extraction with water were determined by a combination of sieve analysis and particle sizing under the microscope. Results for the coarsest and finest ashes are shown in Figs. 4.5 and 4.6, together with one of intermediate fineness. Fig. 4.5 shows that plots of log diameter versus probability weight percent are very nearly linear, implying a Gaussian (normal) distribution about the mass median diameter. The mass median diameters were 46, 23, and 9.6 μm for these samples. The particle size distributions of these same samples are replotted in Fig. 4.6 for comparison with a typical portland cement. This cement had a Blaine specific surface area of 370 m^2/kg or 1166×10^3 m^2/m^3 and a mass median diameter of about 13 μm as determined at the PCA Laboratories with a modern automated instrument (SediGraph™ Particle Size Analyzer), using samples dispersed in a light petroleum distillate. Values of the specific surface areas of the three fly ashes determined by an air permeability method similar to the Blaine method are also shown. To be meaningful, such comparisons between portland cement and fly ash and between different fly ashes (for which the densities are different) should be made in terms of the specific surface in m^2/m^3 rather than m^2/kg. As would be expected on the basis of these values, the portland cement curve lies between the two fly ash curves with higher and lower values (1887 and 1015×10^3 m^2/m^3) over most of the particle size range. There is

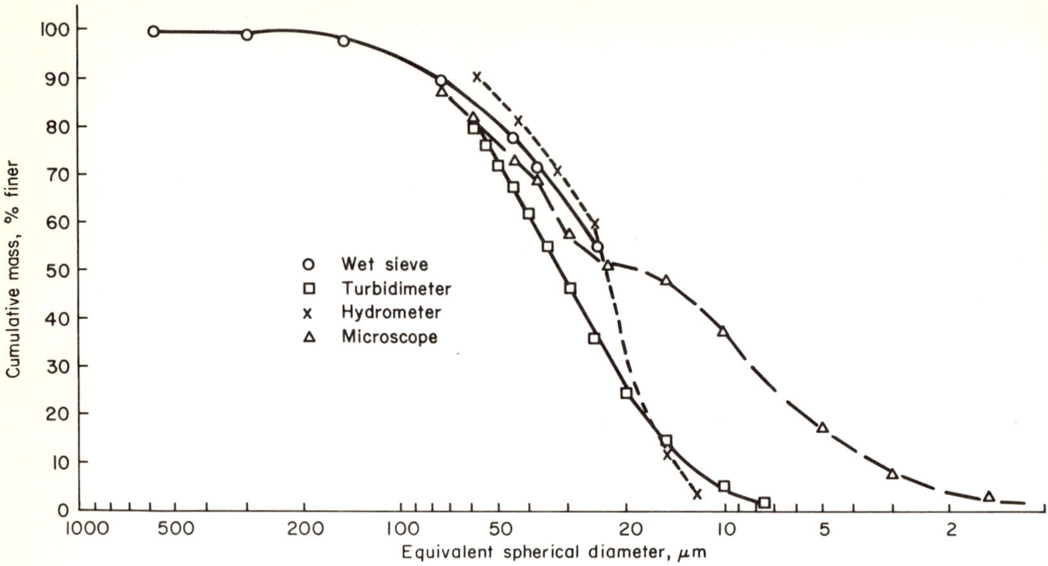

Fig. 4.3 Particle size distribution of a mechanically collected fly ash as determined by four different methods. Blaine fineness = 116.7 m²/kg. (Ref. 121)

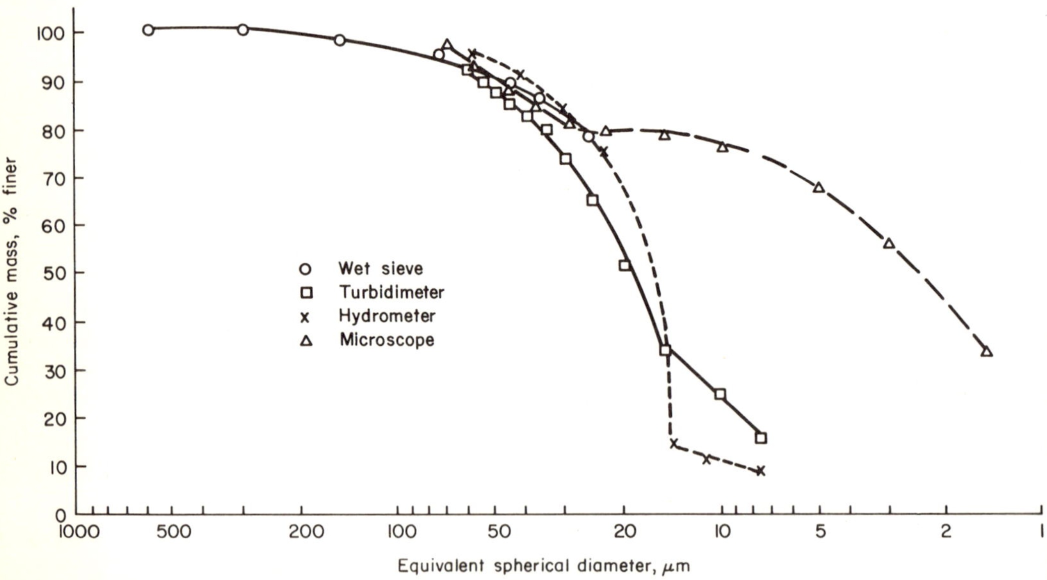

Fig. 4.4 Particle size distribution of an electrostatically precipitated fly ash determined by four different methods. Blaine fineness = 348.0 m²/kg. (Ref. 121)

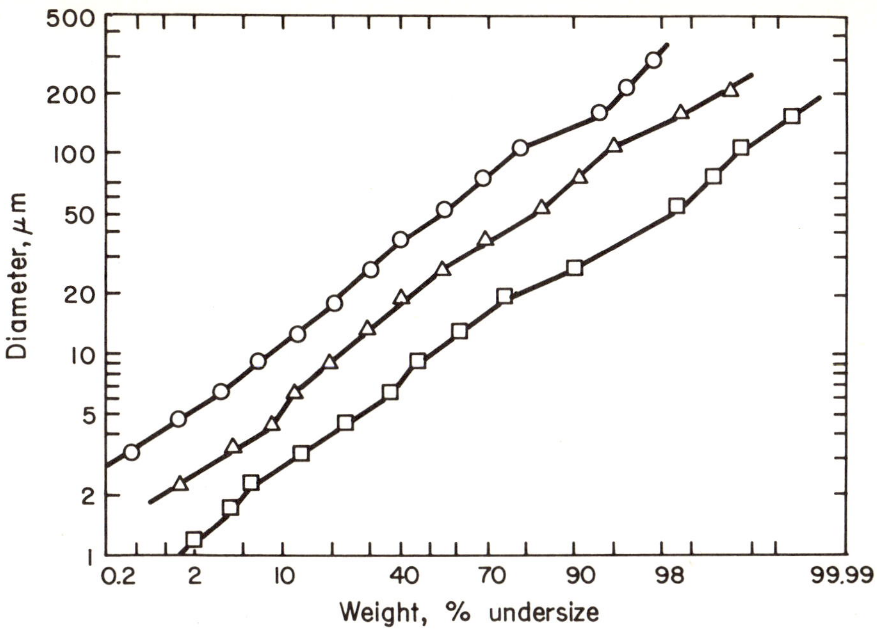

Fig. 4.5 Particle size distributions of typical electrostatically precipitated fly ashes of fine, intermediate, and coarse finenesses. (Ref. 123)

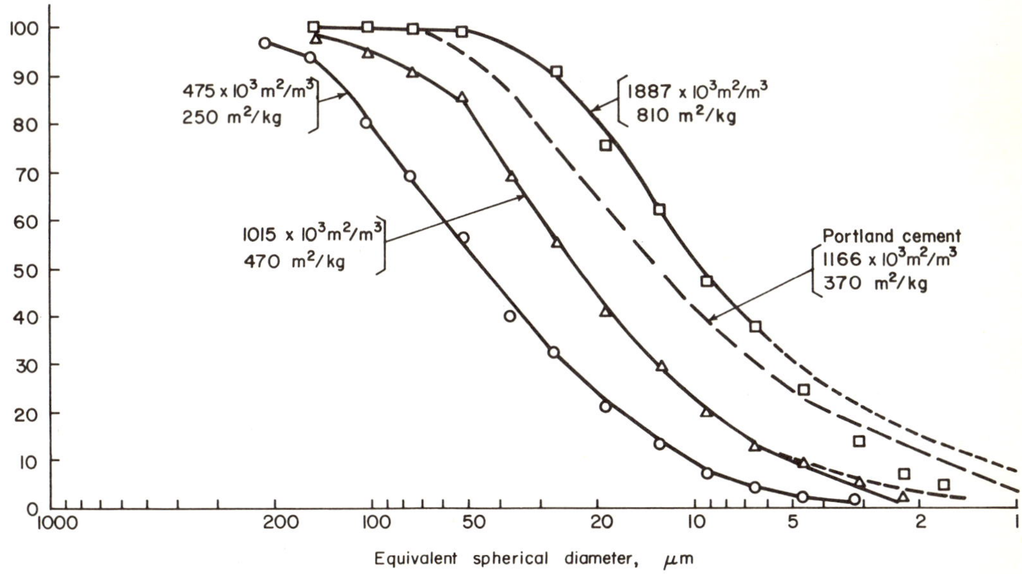

Fig. 4.6 Comparison of particle size distributions of three typical electrostatically precipitated fly ashes with an ordinary portland cement. Finenesses (air permeability method) are shown in both m^2/m^3 and m^2/kg. Fly ash data from Fig. 4.5. (Ref. 123)

some discrepancy in that the fine fly ash curve crosses below the cement curve at about 4 μm.

This discrepancy illustrates again the common problem of systematic errors in the particle size distribution data at the finest particle sizes. The microscopically determined fines percentages are not nearly high enough to account for the air permeability values of the specific surface areas. In all three cases, the specific surface areas calculated from the particle size distributions (228, 454, and 979×10^3 m^2/m^3) are only about one-half of the air permeability values. These facts strongly suggest that, for particle sizes less than about 5 μm, the weight percentages were systematically underestimated and that the true values may have been as indicated by the short-dashed curves. Nevertheless, these data seem quite reliable for sizes above about 7 μm, whereas Figs. 4.3 and 4.4 showed serious discrepancies below about 25 μm.

Ravina[147] determined the fineness of eight samples of low-carbon bituminous coal fly ash collected in different sections of the electrostatic precipitators in a single power station. Finenesses by the air permeability method ranged from 260 to 658 m^2/kg. Since the loss on ignition values ranged up to only 3.8%, the carbon contents were probably low enough not to affect these specific surface areas very much. For these ashes from one source, the amount retained on a No. 325 (45-μm) sieve correlated rather well with the air permeability specific surface areas. Sieve analysis data down to 37 μm were given for each fly ash. Below 30 μm, particle size distributions were determined by sedimentation using the Andreasen pipette method (BS 3046: Part 2).

When fly ashes from different sources are compared, the correlation between the amounts of particles finer than 40 μm and their specific surface areas is poor, as shown in a study of 341 samples of (Class F-like) fly ashes from 14 different power stations[148] and in other studies.[141,143]

As noted earlier, the specific surface area depends mainly on the amounts of the finest particles. Markowski et al.[149] found that there is a sharp peak in the particle size distribution curve at a mass median diameter near 0.16 μm. Modern electrical aerosol analyzers were used to measure particles finer than 1.0 μm used at both the inlet and exit of the electrostatic precipitator on a modern boiler burning pulverized subbituminous Western coal. Complex data analysis was required to extract this information from the noise-ridden raw data, and the results were checked by SEM. The submicron peak, clearly separate from the main particle size distribution curve, contained about 1.5% of the total fly ash. About 90% of the submicron mode was trapped by the ESP; in other installations greater percentages were precipitated. Laboratory studies have since shown that the submicron percentage increases with the combustion temperature.[150] This fine fraction is believed to be formed by condensation from the vapor phase, as discussed in "Combustion and Ash Formation" in Chapter 2, but is not usually detected by methods ordinarily available in the cement and concrete industries.

The literature cited has shown large discrepancies between particle size distributions of fly ashes determined by different methods. Since the densities, refractive indices, and resistivities of fly ash particles vary considerably, even modern automated sedimentation, laserlight scattering, or particle counter instruments cannot be expected to yield highly accurate results for fly ash unless special procedures are used or corrections are made. However, useful results have been obtained by Montgomery[151] and at the PCA Laboratories using a SediGraph™ Particle Size Analyzer. PCA results for three U.S. fly ashes are shown in Fig. 4.7. These ashes had Blaine specific surface areas of 241, 324, and 604m²/kg and mass median diameters of only 16, 11, and 8 μm, respectively, lower than those obtained by Watt and Thorne using other methods for ashes of higher specific surface areas. Such discrepancies need to be resolved.

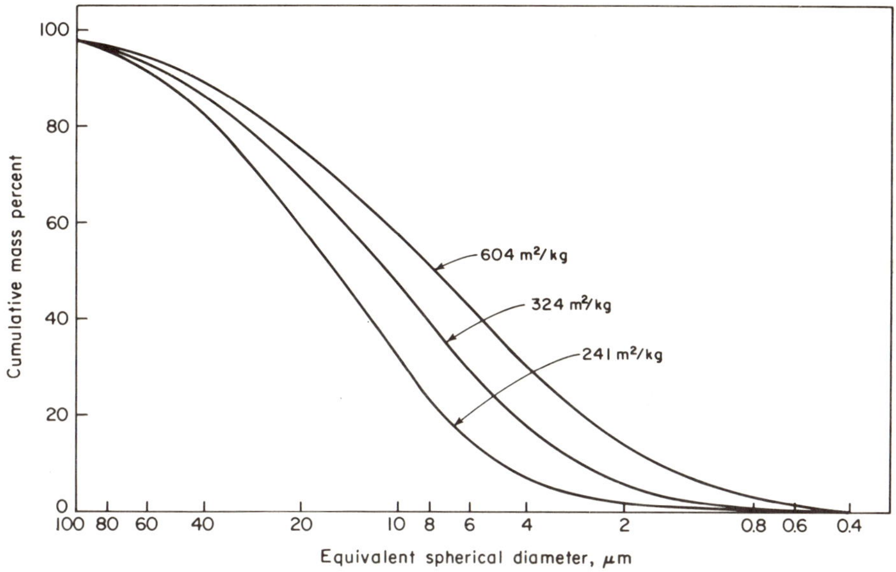

Fig. 4.7 Particle size distributions of three U.S. fly ashes obtained by a modern automated instrument (SediGraph™ particle size analyzer).

Variability of Fly Ash Properties

Variations in coal composition, combustion conditions, ash collection systems, and other variables can have large effects on the composition and fineness of fly ashes. Even fly ash from a single source may vary greatly over relatively short (daily) time intervals.

Power plants operating under stable load conditions tend to produce more uniform fly ash than when under variable load. Although the inorganic oxide compositions may not vary greatly, loss on ignition (LOI) and fineness do. Variations of LOI from 4 to 18% have been reported to result from loads ranging from 100 to 60% of base-load conditions;[152] corresponding changes in Blaine specific surface area were from 320 to 570 m²/kg. From the preceding discussion of fineness it appears that these variations may result mainly from differences in carbon content, reflecting more or less complete combustion conditions, rather than from changes in the fineness of the glassy particles.

Results reported by Jarrige[31B] for variations in the fly ash properties collected at a single power station are shown in Fig. 4.8. Density data as well as Blaine fineness and sieve fractions were obtained. Substantial variations, showing consistent relationships between these variables, occurred on a daily basis. Berry[41B] presented similar results from several other studies.

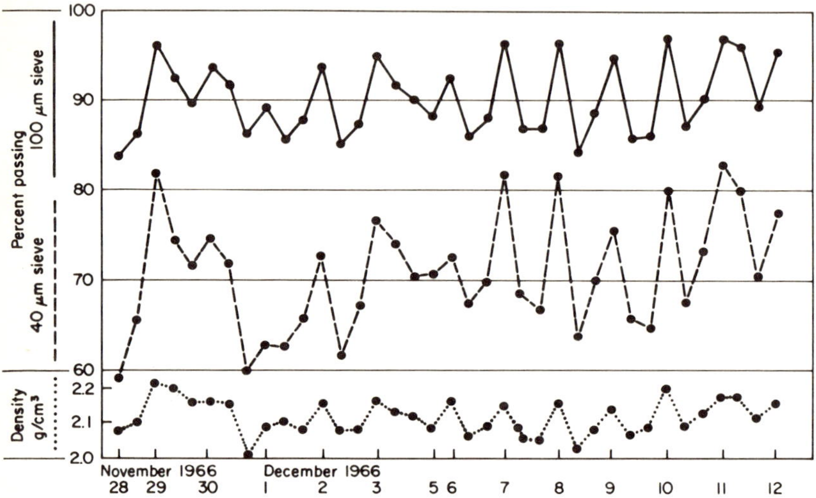

1 cm²/g = 0.1 m²/kg

a. Variations in particle size and density

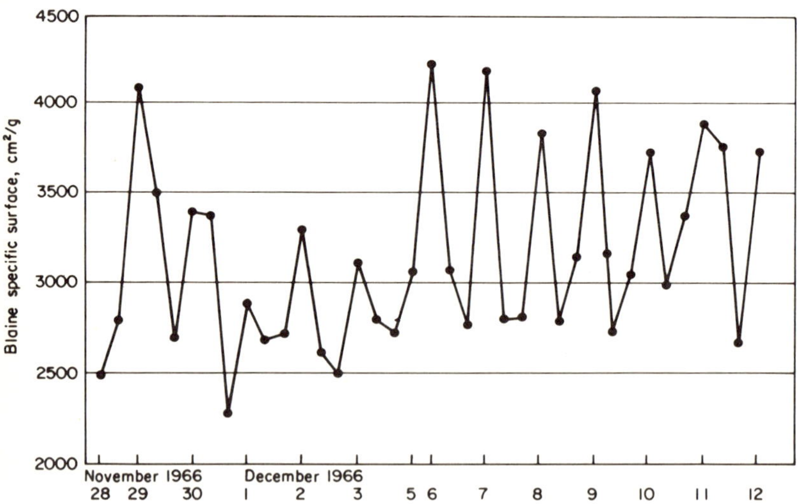

1 g/cm³ = 1 Mg/m³

b. Variations in Blaine surface area

Fig. 4.8 Daily variation in properties of fly ash produced by a generating station under continuous operation. From Jarrige. (Ref. 31B, in Ref. 41B)

Although the cited data indicate very substantial variations in the properties of these samples of fly ashes, there is little indication that any of them would have failed to meet existing ASTM specifications for uniformity (see Table 1.1). Nevertheless, if these materials were used in blended cements and concretes without compensating adjustments, these variations would certainly be reflected in variations in the properties of the concretes.

A systematic study of the effects of variations of properties of fly ashes on concrete strengths has been reported by Dhir et al.[153] using fly ashes from four sources sampled at weekly intervals. Tests were conducted according to British Standard BS3892 for pulverized fuel ash. Concretes were proportioned for 30-mm to 60-mm slump and three different nominal 28-day strengths. The variation in water demand of different samples was accommodated by adjusting the aggregate contents at the mixer for constant slump. Results for the ash showing the largest variations are given in Fig. 4.9 for concretes cured in water at 40° C after demolding. Variations in chemical composition were smaller. These results confirm many other studies that have shown that strengths vary with fly ash fineness, specific surface area, loss on ignition, pozzolanicity index, and water demand. This curing condition yielded the largest number of significant correlations (at the 5% level) with these characteristics, but the correlations were significant only for fineness (45-μm residue and specific surface) and pozzolanicity index for all four ashes. For the results in Fig. 4.9, all five characteristics correlated significantly with strength.

A major concern is that the most critical tests (e.g., pozzolanic activity) require some time to perform. If acceptance of a fly ash is based on such tests, these variations may make adequate sampling and testing to assure quality concrete impractical. A recent discussion by Ashby[154] calls attention to the fact that some of the older published data were obtained with fly ashes that were more variable than some generated in modern power stations. On the other hand, there are also some new and more variable coal sources being used in North America. In one investigation of the properties of Class C fly ashes obtained from Western subbituminous coals, Joshi and Ward[155] found that the rate of strength development was extremely variable and not directly related to the fundamental chemical or physical properties of the ashes. Greatest variability is to be expected in loss on ignition and fineness, and most sources offering fly ash for sale to meet ASTM Specification C618 monitor these two characteristics on a more frequent basis than specified in ASTM C311, often daily.[83]

The variability of Class C fly ashes from five sources in Texas was studied in detail by McKerall and Ledbetter.[81] A wide range of chemical and physical characteristics of the Class C fly ashes was found both between different sources and, to a lesser extent, within a given source over time. ASTM C618 tests for uniformity by No. 325 sieve residue showed about 20% of the 143 samples from two sources "potentially" failed. The "Acid Temperature Rise" test (or "Calcium Oxide Heat Evolution Test") (see "Standard Specifications and Test Methods" in Chapter 1) was also evaluated in interlaboratory testing. It was concluded that this simple test is repeatable, accurate, and rapid (10 minutes). Butler[80] has suggested that testing frequencies in ASTM C311 be modified. He prefers three levels of quality assurance with test frequencies of (1) daily or every 100 tons, (2) monthly composite, and (3) new source evaluation and annual reassessment. The daily monitoring would include tests for fineness, loss on ignition, and, for Class C fly ashes, acid temperature rise. He suggests that these simple daily tests should normally suffice for quality control, more comprehensive testing being required only when indicated by deviations or for annual reassessment.

However, the variability to be expected from any single power plant depends strongly on the variations in operating loads. If these vary greatly with time of day, large variations within one day will occur, especially for specific surface area, as shown in Fig. 4.8. To understand more fully how these variations affect the properties of concrete, we must examine the properties of fresh fly ash-cement mixtures and the pozzolanic and cement hydration interactions.

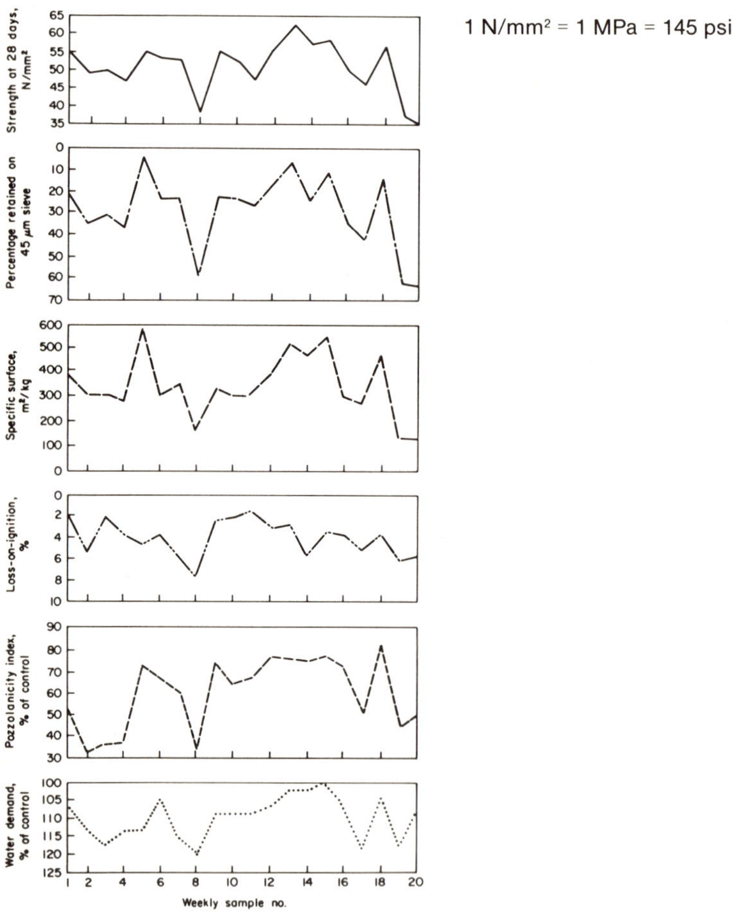

Fig. 4.9 Comparison of ash characteristics variability with variation in strength of concretes made with 41% fly ash (by weight of cement plus fly ash) and cured at 40°C. (Ref. 153)

Chapter 5
Fresh Fly Ash-Cement-Water Mixtures

Among the most important characteristics of fly ashes are those that affect the water requirement for flow, workability, bleeding, and setting characteristics of the fresh cement pastes in concretes. Replacement of a part of the cement with fly ash may either increase or decrease the water requirement. Similarly, the bleeding of concrete may be either increased or decreased.(3) The interactions with commonly used admixtures such as air-entraining agents, water-reducing admixtures, accelerators, and retarders are also important. In addition to affecting the properties of fresh concretes, these characteristics influence the water-cement ratio used and the properties of the hardened concrete.

The effect of variations in cement composition on the early hydration reactions, which can affect the rheology of the fresh mixtures, will be considered in Chapter 8. In this Chapter attention is focused on cements with well-controlled early hydration behavior.

Microstructure and Rheology

Many investigators have reported that partial replacement of portland cement by some fly ashes in mortars and concretes reduces their water requirements for flow. This phenomenon is commonly believed to be a result of the spherical shape and smooth surface of many of the fly ash particles, by analogy with the proven effects of shape of macroscopic aggregate particles. Although some data have been obtained which seem to support this belief,(3) this explanation now appears to be entirely too simplistic.

Yamazaki(156) made a systematic study of the effect of particle shapes of several fly ashes and other powdered minerals on the water requirement for flow and the bleeding rates of fly ash slurries. It was found that water reduction increased with fineness of the fly ash and that the finer fly ashes had a larger proportion of smooth spherical particles. He then concluded that the smooth spherical shape was the most important reason for the improved workability of concrete due to fly ash. However, if the cement and fly ash particles are normally in a flocculent state, it does not seem likely that the individual particle shapes can have much effect on the flow. The flow consists of the motion of flocs of different sizes and shapes containing pockets of entrapped water. As shown in studies of the effect of water-reducing admixtures, the major water reduction effect is that of breakdown of the floc structure. Yamazaki did recognize the importance of breakdown of the agglomerates of fly ash particles formed in the dry state; when mixed in concrete, these particles of fly ash are difficult to disperse unless first vigorously mixed as a fly ash-water slurry.

The microstructure and rheological behavior of aqueous suspensions of cement and fly ash particles are complex.(157) Such suspensions normally have a flocculent structure, as shown in Fig. 5.1.(3) This structure of particles in various states of agglomeration results from the interparticle electrostatic forces.

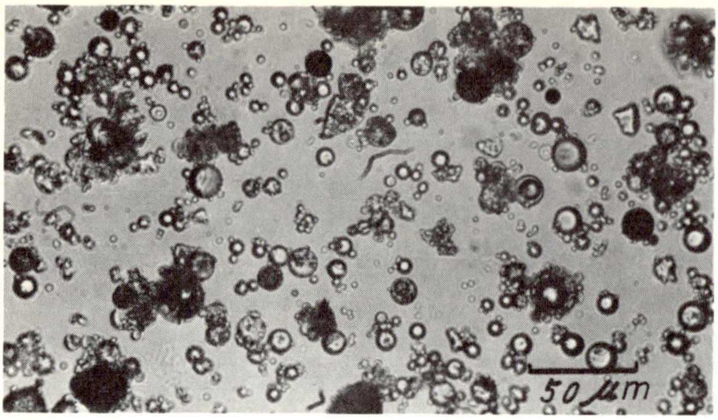

a. Fly ash particles dispersed in pure water.

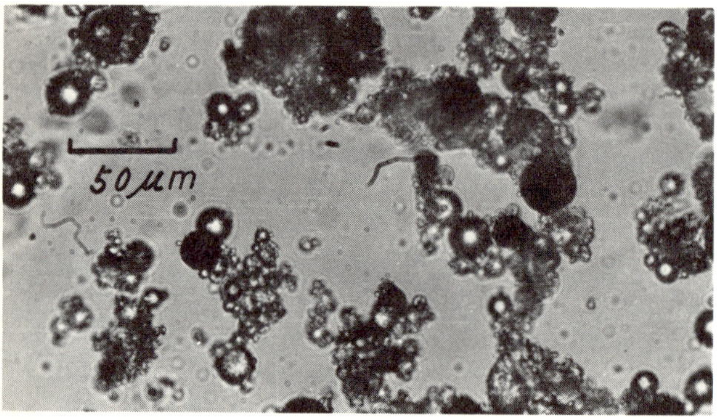

b. Flocculated fly ash particles in supernatent liquid of cement paste.

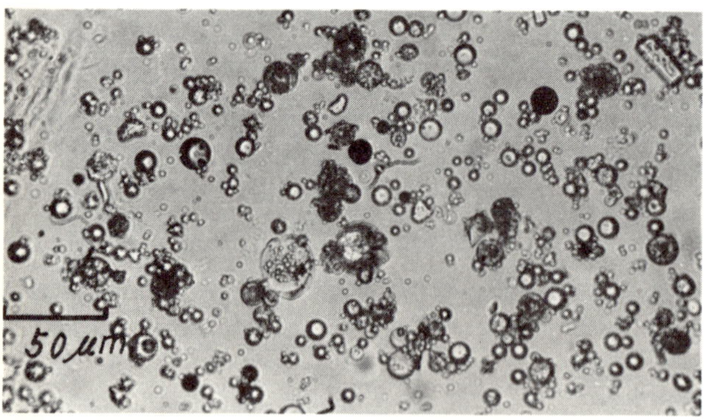

c. Fly ash particles in supernatant liquid of cement paste containing suitable amount of cement-dispersing agent.

Fig. 5.1 Flocculation and dispersion of fly ash particles in water and supernatant liquid. (Ref. 3)

Bombled[158] pointed out that the silicate and aluminate phases in portland cements immersed in water have surface charges of opposite signs (silicates are negative, aluminates positive). The fine particles are mostly monomineralic, so that they form floc structures of weakly bonded particles of opposite signs alternating in the floc. By contrast, particles of the same sign (such as calcium aluminate cements) are dispersed by mutual repulsion forces, and suspensions of such particles flow more easily. The magnitude of these forces also depends on the concentration and distribution of ions in solution. For pH values far enough from the value for zero charge, the repulsive electrostatic forces overcome the attractive Van der Waals forces, and the suspensions do not flocculate.[159] The particular fly ash shown in Fig. 5.1 was not strongly flocculated in pure water, but this micrograph indicates that it became so when mixed in the supernatant liquid from a cement paste, which contained the soluble constituents from both the cement and the fly ash. Powers found that limewater was necessary for some powdered minerals to produce a normal flocculent structure.[160] Also shown in Fig. 5.1(c) is the effect of a cement-dispersing agent or water-reducing admixture. Such admixtures neutralize the charges on the particle surfaces and reduce interparticle attractive forces. With sufficient admixture dosage, the flocculent structure is destroyed so that the particles are completely dispersed in the mixing water.

The flocculent structure of cement pastes may also be broken down mechanically during mixing, but there is a counteracting tendency for reformation of the structure. There is a shear-thinning process that takes place during mixing, and thickening takes place while the paste is at rest. Such behavior is termed thixotropic. In addition, there are mechanical particle interference effects as the larger particles and flocs move past one another during flow. At very low water contents or at very high shear rates[161,162] this results in shear-thickening, dilatant behavior. The average thickness of the water films between particles is directly proportional to the water-cement ratio and inversely proportional to the specific surface area of the cement, so that the resistance to flow of the paste increases with the fineness of the cement and decreases with water content.

Ordinary cement pastes require a finite yield stress before flow begins because of the interparticle forces. They are approximately Bingham plastic solids with very low values for the yield stress (up to 1,000 Pa or 0.145 psi) and plastic viscosities ranging from 5 to 50 mPa·s (centipoise). These values depend on cement fineness and water-cement ratio and are greatly reduced if water-reducing admixtures are used; if the flocculent structure is thereby destroyed, the yield value is reduced to essentially zero. The suspension is then an approximately Newtonian fluid with much simpler rheological behavior. It is apparent that the flow properties of cement-fly ash-water mixtures must depend on much more than the smooth shapes of fly ash particles.

Rheological Properties of Fresh Pastes

In an experimental study of fly ash-cement pastes, Ivanov and Zacharieva[163] obtained approximately Binghamian rate of shear strain versus shear stress curves using a coaxial cylinders viscometer. Water-cement ratios (w/c) ranged from 0.45 to 0.75 and fly ash contents from 0 to 100%. Two fly ashes of similar chemical composition but different finenesses (465 and 515 m^2/kg) were employed. Results obtained at 0.6 w/c are shown in Fig. 5.2. These results show progressive increase in the yield stress and plastic viscosity (the reciprocal of the slope of the linear portion) with increasing fly ash contents, except for the lower fineness fly ash, which was slightly less viscous in the 100% fly ash paste than at 65% fly ash. (The water-cement ratio indicated for the 100% fly ash paste is apparently the water-fly ash ratio.) Yield values obtained for similar mixtures at four w/c's are given in Table 5.1. Values of the plastic viscosity increased similarly with fly ash content.

In a theoretical study of the effect of fly ash on the workability of cement paste and concrete, Hobbs[142] used equations developed for the effect of aggregate volume concentration on the yield value and plastic viscosity. The Bingham equation for the shear stress τ_{bc} in a paste made with fly ash-blended cement then became

$$\tau_{bc} = \tau_{0p} f + U_p g \gamma_{bc} \qquad (5.1)$$

in which τ_{0p} is the yield stress of the cement paste, U_p is the plastic viscosity of the paste, γ_{bc} is the rate of shear strain, and f and g are functions of the ash volume

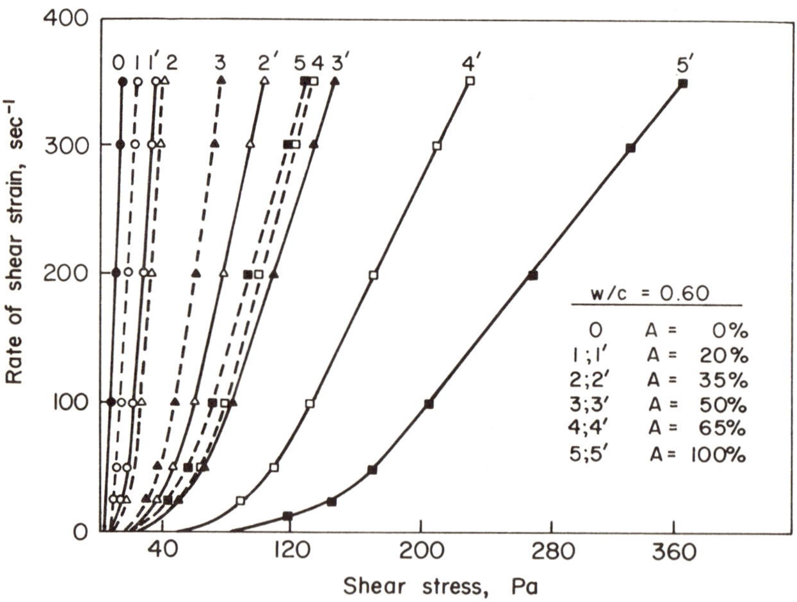

1 Pa = 1 N/m² = 0.145 X 10^{-3} psi

Fig. 5.2 Influence of the fly ash contents (A = % ash) in the cement on the flow of the paste in a coaxial cylinder viscometer. (Pastes with 465 m²/kg ash are labeled 1 . . . 5; pastes with 515 m²/kg ash are labeled 1′ . . . 5′.) (Ref. 163)

Table 5.1 Yield-Stress Values (Pa) of Pastes at Various Fly Ash Contents and Water-Cement Ratios[163]

Fly ash fineness, m^2/kg	Fly ash in blend, %	Water-cement ratio			
		0.45	0.55	0.65	0.70
465	0	46	13	6.2	4.5
	20	58	18	8	6.5
	35	77	35	11	9.5
	50	185	66	20	16
	65	238	89	48	32
	100	160	70	34	26
515	20	77	35	10	8
	35	150	63	37	19
	50	254	80	48	30
	65	—	124	64	47
	100	—	248	110	89

1 Pa = 1 N/m^2 = 0.145 X 10^{-3} psi

concentration, V_a, and the volume concentration for close packing, $V_{a,m}$. Below are the functions f and g:

$$f = 1 + \frac{1.5 V_{a,m} V_a + V_a^2}{(V_{a,m} - V_a)} \tag{5.2}$$

$$g = 1 + \frac{2.5 V_{a,m} V_a}{(1 - V_a)(V_{a,m} - V_a)} \tag{5.3}$$

$V_{a,m}$ varies from 0.47 to 0.71 for fly ashes ranging in bulk density from 1.12 to 1.49 g/cm^3 and in density from 2.1 to 2.4 g/cm^3. The value of f (or g) increases from 1.0 to 2.0 (or 3.5) as fly ash contents increase from 0 to maximum values.

Hobbs used these equations with Vom Berg's data[161] for τ_{0p} and U_p to calculate the corresponding values for blended cement pastes. These are shown in Figs. 5.3 and 5.4 as functions of the true water-cement ratio (w/OPC). If the true water-cement ratio is kept constant and fly ash is added, the blend becomes stiffer because of the increased concentration of total solids. Extra water must then be added to achieve the same flow, and the true water-cement ratio is increased, although the water-blended cement ratio may be lower than for the cement paste.

To test Eqs. 5.2 and 5.3, the data of Ivanov and Zacharieva were used by the writer to obtain the curves plotted in Figs. 5.5 and 5.6.[164] The experimental data are seen to be qualitatively very similar to those obtained using Hobbs' equations. (Points A-E in Fig. 5.5 are not data points but are mixtures that will be referred to in the discussion of Table 5.2.)To check the equations quantitatively, the yield values and plastic viscosities of the blended cements were calculated from the cement paste values assuming the same typical values for $V_{a,m}$ (0.6) and fly ash density (2.2 g/cm^3) used by Hobbs to calculate the curves in Figs. 5.3 and

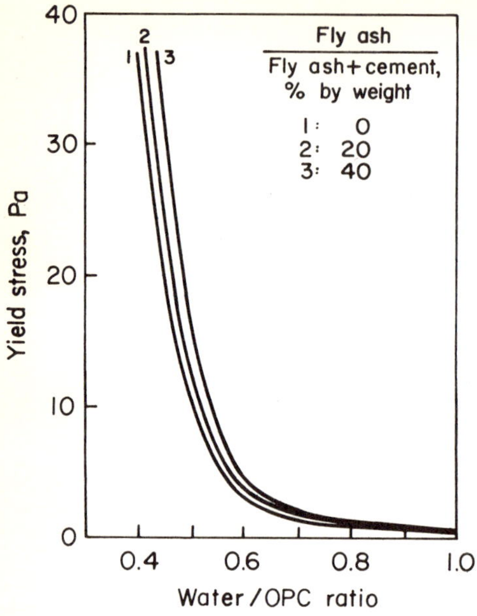

Fig. 5.3 Dependence of predicted yield stress of blended cement paste upon the water/OPC ratio. (Ref. 142)

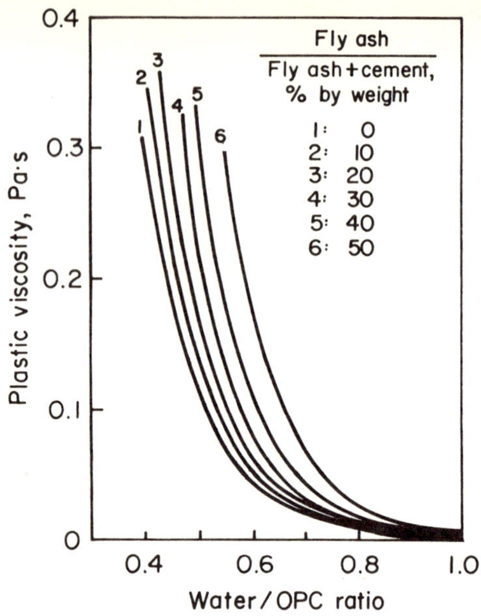

Fig. 5.4 Dependence of predicted plastic viscosity of blended cement paste upon the water/OPC ratio. (Ref. 142)

5.4. The results shown in Figs. 5.7 and 5.8 are for the ratios of the experimental to the calculated values as functions of the fly ash contents. These results show that the experimental values of yield stress and plastic viscosity are always higher than those calculated from Hobbs' equations, becoming 200 to 300% higher at high fly ash contents. For fly ash replacements of less than 20%, the calculated values are good approximations, being within 20% of the experimental values. Expected variations of the assumed values of $V_{a,m}$ and fly ash density do not change these curves appreciably.

The curves in Fig. 5.5 can also be used to examine the effects of additions of fly ash and variations in w/c in fly ash-cement-water mixtures in another way. The 0.5 w/c cement paste at point A may be modified by additions of the same amount of fly ash (e.g., addition of 25% fly ash by weight of cement or replacement of 20% of the weight of cement) with corresponding different effects on the rheological properties, paste volume, and paste porosity (taking the water volume fraction as the porosity). Four examples of paste blends with 25% fly additions are given in Table 5.2 and indicated in Fig. 5.5 by points B, C, D, and E. Also given in Table 5.2 for comparison with each fly ash mixture are two additional cement paste mixtures designated BCV, BCY, and so forth, to indicate that they are equivalent to fly ash mixture B in paste volume or yield value.

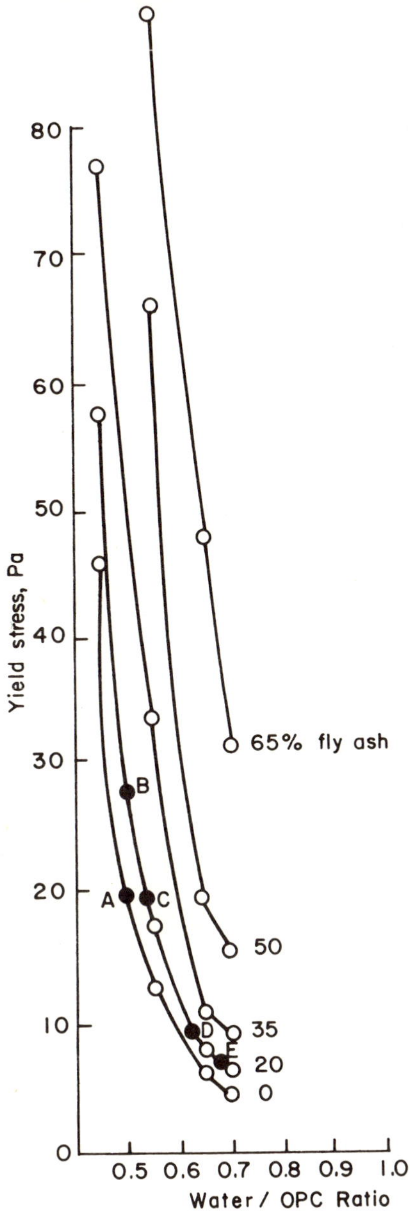

1 Pa = 1 N/m² = 0.145 X 10⁻³ psi

Fig. 5.5 Experimental values of yield stress of cement pastes containing various amounts of fly ash of 465 m^2/kg. Solid points are not data points but are compositions used in Table 5.2. (Ref. 164)

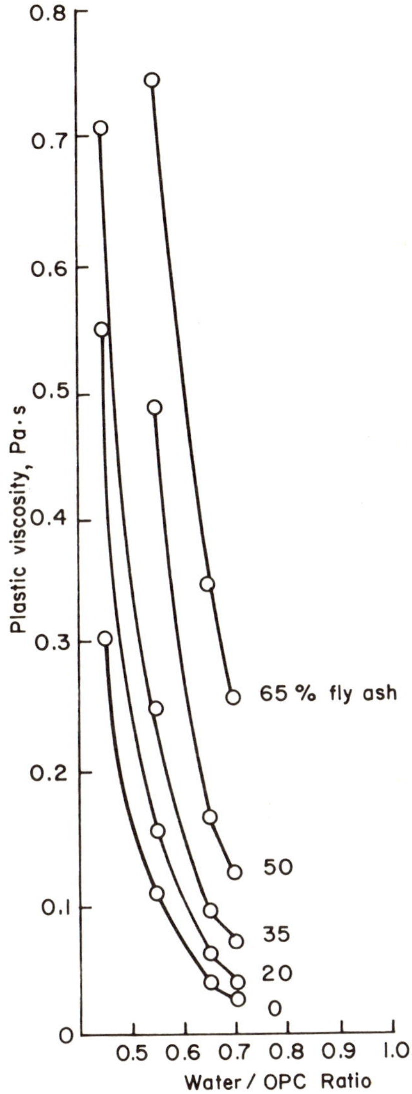

1 Pa·s = 1 centipoise

Fig. 5.6 Experimental values of plastic viscosities of cement pastes containing various amounts of fly ash of 465 m^2/kg. (Ref. 164)

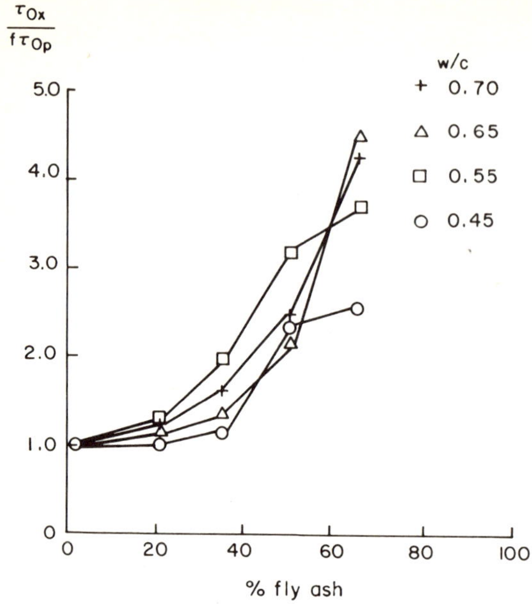

Fig. 5.7 Test of Hobbs's equation for dependence of yield stress on fly ash content. τ_{0x} is the experimental value; $f\tau_{0p}$ is the theoretical value. (Ref. 164)

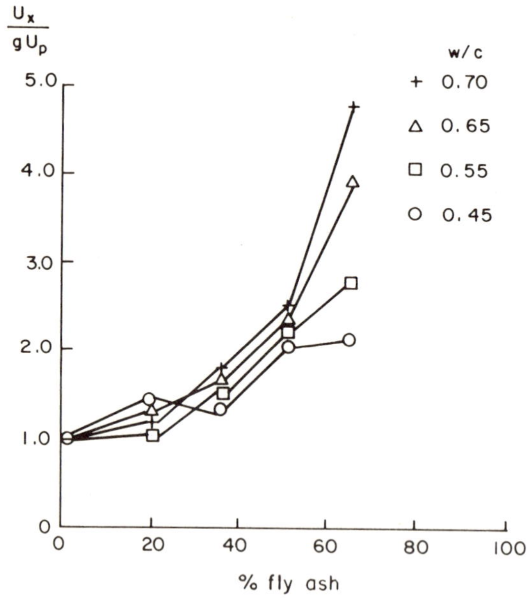

Fig. 5.8 Test of Hobbs's equation for dependence of the plastic viscosity on fly ash content. U_x is the experimental value; gU_p is the theoretical value. (Ref. 164)

Table 5.2 Effects of 25% Additions of Fly Ash to 0.5 Water-Cement Ratio Paste and Properties of Equivalent Cement Pastes*

Mix	Condition	Additions, kg/kg cement		Effects		
		Fly ash	Water	Yield stress, Pa	Paste volume increase, %	Paste porosity
A	Cement paste, w/c = 0.5	0	0	20	0	0.61
B	No water added	0.25	0	28 (+40%)	14	0.54
BCV	Equal paste volume	0	0.114	8	14	0.66
BCY	Equal yield value	0	-0.03	28	-3.7	0.60
C	Water added for equal yield value	0.25	0.040 (8%)	20	19	0.56
CCV	Equal paste volume	0	0.154	6	19	0.67
CCY	Equal yield value	0	0	20	0	0.61
X	[see text]	0.25	0.091 (18%)	13	25	0.58
D	Water added for equal water/solids weight ratio	0.25	0.125 (25%)	9.5 (-53%)	29	0.59
DCV	Equal paste volume	0	0.239	3.5	29	0.70
DCY	Equal yield value	0	0.09	9.5	11	0.65
E	Water added for equal water/solids volume ratio	0.25	0.18 (36%)	7 (-65%)	36	0.61
ECV	Equal paste volume	0	0.293	2.5	36	0.71
ECY	Equal yield value	0	0.13	7	16	0.66

*Based on fig. 5.5 derived from data by Ivanov and Zacharieva [163] and equations by Hobbs.[142] (Ref. 164)
1 Pa = 1 N/m^2 = 0.145 × 10^{-3} psi

The fly ash mixtures are listed in order of increasing water contents from B, which has no additional water, to E, which has 0.18 kg water/ kg cement, or 36% added water. The composition of mixture X (see Table 5.2) is the same as that obtained by replacing 20% of the cement in mixture A with an equal weight of fly ash and reducing the water content to maintain constant paste volume. Paste B with no added water is the only paste with a yield value higher than that of pure cement paste.

It is interesting to note that, although some water is added to the fly ash mixtures C, D, and E, the porosity values of these pastes are less than or equal to those of the cement paste, the paste volumes are greater, and the yield values are less than or equal to those of the cement paste. Since increasing the paste volume increases the workability of concrete unless the paste is much stiffer, this may be the most important factor with respect to improved workability of concrete. Even mixture B may improve concrete workability because of the increase of paste volume, despite the higher yield value.

It should also be noted that, when water is added instead of fly ash to obtain the same yield value (mixes DCY and ECY), the paste volumes are much less than with the fly ash, while the porosities are higher. Also, when water is added instead of fly ash to obtain the same paste volumes, the yield values are lower in each case, but the w/c and porosities are much higher. The addition of fly ash with a little water instead of just water results in reduced porosities as well as increased paste volumes, reduced yield values, or both, for all of these blends.

Water Requirement of Mortars and Concretes

Prior to the cited work by Hobbs[142] on the workability of concrete, Powers[160] had developed a fundamental concrete proportioning procedure which included the use of fly ashes and other mineral admixtures. He showed that if the maximum particle size of the fly ash was not excessive, cement paste can be replaced by an equal volume of fly ash paste of equal consistency without affecting the workability of the concrete. If the water requirement of the fly ash paste is low, the result is a higher solids content for the concretes.

In a comprehensive statistical analysis of data contained in several major experimental studies, Minnick, Webster, and Purdy[141] showed statistically significant correlations between the water requirement of mortars and certain characteristics of the fly ashes. The most consistent correlations with the water requirement were obtained for loss on ignition (LOI), or carbon content, and No. 325 sieve residue. The effect of the LOI should be expected because of absorption of water by porous carbon particles, and the effect on No. 325 sieve residue should be expected because of particle interference effects.

Analyses were also done with a restricted set of data thought to be more representative of modern fly ashes. When the No. 325 sieve residue (325S), loss on ignition, and their product were held within specified ranges (12% max. LOI, 20% max. 325S, and 150 max. for the product or multiple factor), the relative water requirement (W.R.) as the percentage of the control mixtures without fly ash was found to be

$$\text{W.R.} = 92.6 + 0.086(\text{LOI} \times 325\text{S}) \qquad (5.4)$$

The average deviations from this equation were less than 3% of the values. For the restricted set of data, a multiple factor of 150 corresponded to a water requirement of about 105% of the controls, which is the ASTM C618 maximum. The slope in this linear regression equation decreased when the coarser fly ashes with more than 20% retained on the No. 325 sieve (up to 34%) were included; hence, the maximum value for the multiple factor adopted in ASTM C618 as an optional physical requirement for Class F fly ash was 255, rather than 150. The 1986 ASTM C595 and C618 specification limits are shown in Fig. 5.9, along with these relationships between the LOI, No. 325 sieve residue, and multiple factor, assuming that Eq. 5.4 is valid.[91] As it has turned out, that restricted set of data is *not* very representative of modern U.S. fly ashes. There are still many fly ashes with No. 325 sieve residues well over 20%. The LOI values, however, are now often

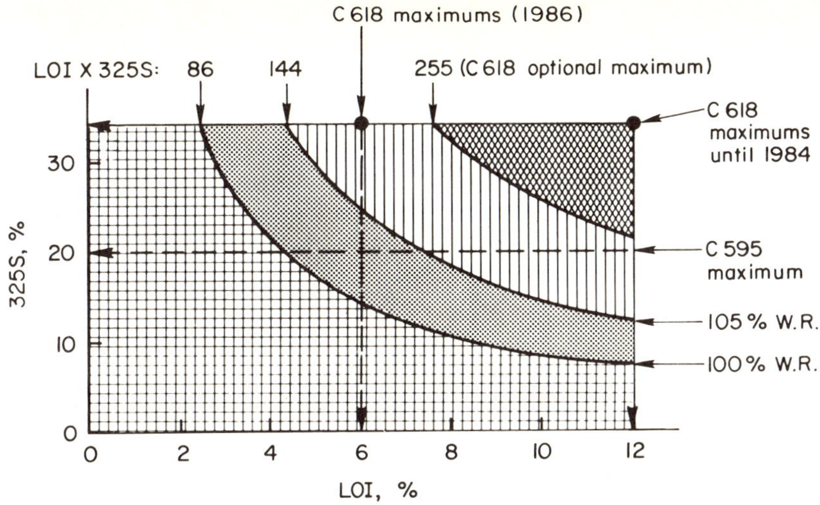

Fig. 5.9 ASTM C618 fly ash maximums for LOI, 325S, LOI × 325S, and the zones corresponding to water requirements of mortars given by Eq. 5.4. The C618 Class F fly ash maximum LOI was lowered to 6% in 1984. (Ref. 91)

well below 5%, making the effect of the No. 325 sieve residue all the more important in the mortar flow test prescribed in ASTM C311.

A fly ash with a high percentage of +325 material will have a high water requirement in the mortar flow test because the volume of paste is not large enough to avoid particle interference effects. This is because of the fixed proportions of fly ash, cement, and sand (1 part cement to 2.75 parts sand, with 35% by volume of the cement replaced with fly ash). In concrete, fixed proportions are not required, and the coarse fraction of the fly ash can be regarded as sand, with the fly ash and sand contents adjusted accordingly. If the proportions in the mortar flow test were adjusted to include the amount of +325 mesh material in the fly ash with the sand, the results should provide more useful information concerning the true water-reducing characteristics of the finer fraction of the fly ashes in concrete.[165]

Water-Reducing Admixtures

The effect of high-range water-reducing admixtures on the water requirements of mortars containing fly ashes was examined in a recent study by Stuart et al.[166] The results also showed that the fly ashes themselves were effective as water-reducing admixtures, although differing considerably in their effectiveness. Four different fly ashes were used with one ordinary portland cement. Two admixtures were used, namely, a sulfonated melamine formaldehyde condensate (melamine-type) and a naphthalene sulfonate formaldehyde condensate (naphthalene-type). ASTM C109 mortars were prepared using a high silica (White Marsh, Maryland) sand with water contents adjusted to yield a flow of 110±5% as determined by flow table tests. The w/c of the plain portland cement mortar was 0.63, which suggests that the sand used had a water requirement much higher than the commonly used Ottawa (Illinois) silica sand. Fly ash replacements of 20, 40, and 60% by volume of cement were used. Because volume replacements were used, the portland cement plus fly ash to sand ratio by weight did not remain constant at 1.00:2.75 as with portland cement mortars; the ratio decreased slightly as the volume of ash increased. Dosages of the organic admixtures were based on the weight of the portland cement alone and were found by test to be optimum at 3.25 and 1.2% for the melamine-type and naphthalene-type, respectively.

Preliminary studies showed that the fly ashes did not change optimum dosages of the admixtures in the mortars; apparently the admixtures were not adsorbed by the fly ash. Physical properties of the fly ashes are given in Table 5.3, and results of the tests for water requirements are shown in Fig. 5.10.

Fig. 5.10a shows that the two finest ashes, D and A, reduced the water requirement to distinct saturation levels with 20% (or less) fly ash and that further additions had little effect. Figs. 5.10b and c indicate that only a certain maximum quantity of water can be eliminated with the organic admixtures and that this is not exceeded with fly ash additions or affected by the differences in water-reducing properties of the ash. When the particles are in the dispersed state, the differences between the ashes, and even the differences between fly ash and cement particles, no longer have any appreciable effect on the water requirements.

The Mechanism of Water Reduction by Fly Ashes

To explain the water-reducing properties of some fly ashes, several alternatives must be considered. We cannot account for the water reduction in the tests by Stuart et al. in terms of the increased paste volume in the mortars, as suggested in the discussion of Table 5.2, because the cement was replaced on an equal volume basis; because the water contents were reduced, the volume of paste was actually reduced with fly ash replacements. In "Microstructure and Rheology" in Chapter 5, the effect of particle shape was discounted for particles of cement fineness in the flocculent state. Data of Fig. 5.10a confirm this reasoning because, if the water reduction resulted from spherical shapes of fly ash particles or from their

Table 5.3 Physical Properties of Four Fly Ashes[166]

Fly ash	Specific gravity	Blaine specific surface, $m^2/m^3 \times 10^{-3}$	Weight Retained, %	
			No. 325 Sieve	No. 200 Sieve
A	2.59	793	7.4	2.5
B	2.32	517	17.7	9.8
C	2.29	448	18.3	8.5
D	2.89	806	7.8	3.3

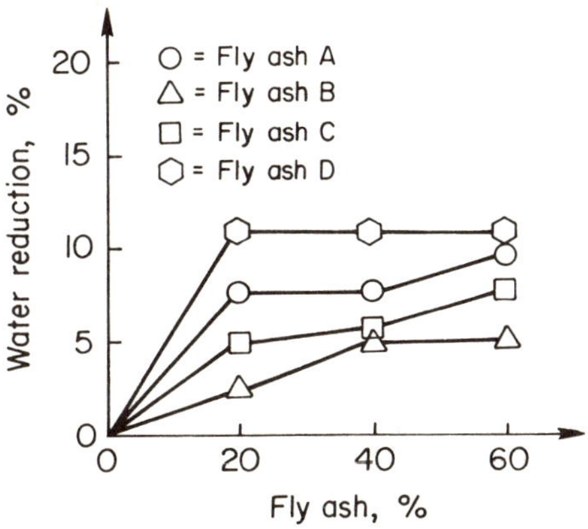

a. No organic admixture

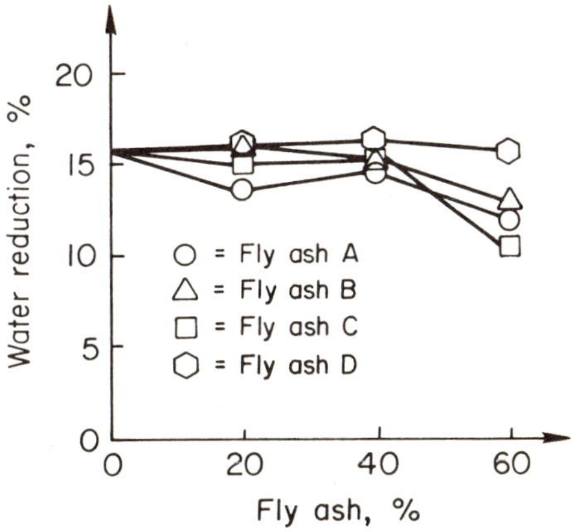

b. Melamine-based admixture

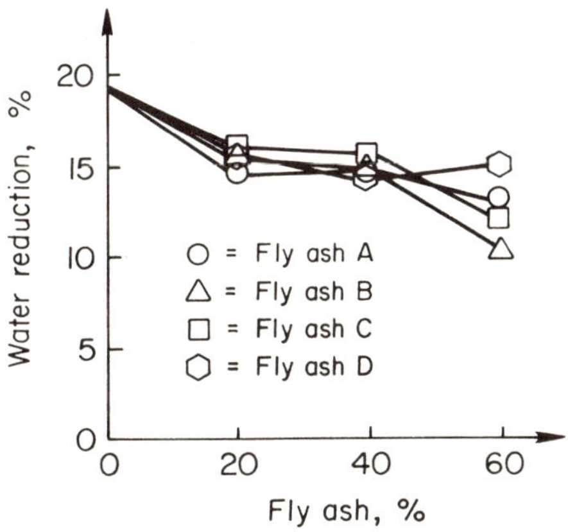

c. Naphthalene-based admixture

Fig. 5.10 Water reduction versus fly ash replacement for portland cement in mortars. (Ref. 166)

lack of chemical reactivity, increasing the fly ash contents would progressively reduce water requirements instead of reaching the distinct saturation levels indicated for ashes A and D. Even in the case of the mortars dispersed with the organic admixtures, the differences between the shapes (or chemical reactivity) of cement particles and fly ash particles apparently had no effect; there was no appreciable variation in water reduction with fly ash content.

What these results do suggest is that the water reduction caused by the fly ash is a result of an adsorption and dispersion process very much like that of the organic water-reducing admixtures.[167,168] The abrupt change in the increase in water reduction at 20% (or less) fly ash for ashes A and D strongly resembles the saturation effect observed with the high-range water-reducing admixtures. This behavior might be explained in either of two ways. The first is that the water reduction effect of the fly ashes may be a result of the release of adsorbed organic compounds from the fly ash. This hypothesis would qualitatively explain the apparent saturation effect shown in Fig. 5.10a with ashes A and D and would also be consistent with the greater effectiveness of the highest specific surface ashes. However, it does not quantitatively fit with the fact that, when the organic admixtures were added to fly ash mixtures at the saturation levels, considerably greater water reduction was achieved. It is also inconsistent with the fact that fly ash particles are themselves flocculated in cement liquor.

These facts suggest another hypothesis: that the water reduction is a result of the adherence of very fine fly ash particles on the cement particles. This same hypothesis was proposed and investigated by Yuan et al.,[167] who attributed slight reductions of viscosities, yield values, water requirements, and transparency of cement-fly ash pastes to adsorption of "microparticles of fly ash" on cement particles. If a large proportion of the fly ash particles are glassy particles with surface charges of the same sign, they will neutralize charges on cement particles of the opposite sign. Fly ash with a (negative) surface charge will come into contact in the course of mixing with the positively-charged aluminate and calcium-rich areas on the surfaces of the calcium silicates. These particles should adhere to those surfaces; in fact, the very fine fly ash particles may be regarded as being "adsorbed" by larger cement particles. If there are enough fine fly ash particles to completely cover the positively-charged areas of individual cement particles, the cement particles will be effectively dispersed and will either flow more easily or have a reduced water requirement for flow.

This hypothesis appears to account for most of the facts concerning the water requirements of the mortars used by Stuart et al.[166] It may also account for the somewhat lower water reduction achieved with the best fly ash (D) than with the organic admixtures. If this mechanism does in fact operate as described, there would be some water trapped in the interstices between the ash particles on the surfaces, increasing the water requirement somewhat above that of the completely dispersed system obtained with the organic admixtures. For the most effective ash (D), the water reduction is about 16–12 = 4% less than that achieved with the organic admixtures at most ash contents, or $0.04 \times 0.63 = 0.025$ kg water/kg cement. A simple calculation indicates that this much water could be retained as the interstitial water in a single layer of "adsorbed" fine fly ash particles.[168] It must be concluded that this alternative hypothesis for the water reduction effect of some fly ashes does approximately account for these experimental results. Yamazaki's conclusion (on the first page of Chapter 5) that the spherical shape was the most important factor was therefore erroneous.

Water requirements of cement-fly ash mixtures involve three other factors that may be important. Organics in the fly ash may also contribute to the dispersion of the cement particles. If the early hydration reactions are not properly controlled, the hydration of the tricalcium aluminate may produce hexagonal phases in the aqueous phase instead of dense coatings of ettringite on the aluminate faces of the particles. Very fine cement particles may also function in the same way as postulated for the fine fly ash particles. Most cements, however, contain very little material as fine as 0.2 μm, and this material may dissolve rapidly.

Air Entrainment

All cement pastes, mortars, and concretes entrap a certain amount of air during mixing, even if an air-entraining (AE) admixture is not used. This is especially true in mixtures with low water to solids ratios. The entrained air increases the voids content and improves the flow properties of the fresh mixtures. This process has been explained in detail by Powers, using his voids ratio diagram,(5B) which showed the advantages of using paste-making materials such as fly ash and entrained air in addition to portland cement and water in such mixtures. By contributing to the fluid volume, the air voids substantially reduce the water requirement for flow of mortar and concrete.

When air-entraining admixtures are used in portland cement pastes, mortars, and concretes to ensure frost resistance, the amount of entrained air can usually be controlled within narrow limits by adjusting the amount of AE admixture. These air voids are spherical bubbles that are stabilized by the reaction product between the AE admixture and the cement liquor. The voids vary in size from microscopic to upwards of about 1 mm and may account for 20% or more of the volume of the paste.

AE Admixture Requirement

In mixtures containing fly ashes, the amount of AE admixture required to produce a given percentage of entrained air has usually been found to be higher, and sometimes much higher, than in comparable mixtures without fly ash.(3,169,170) This high admixture demand affects the water requirement for flow of the fresh mixture and may also affect the durability of the hardened concrete if there is continuing loss of entrained air during mixing and handling of the fresh concrete prior to placement. For these reasons, studies have been made of the effects that fly ashes have on air entrainment and air loss during prolonged mixing, as may occur in a transit mixer.

The curves in Fig. 5.11 show the variations in AE admixture requirements of mortars made without and with four different fly ashes.(169) The properties of the fly ashes used in this study by Meininger are given in Table 5.4. The mixtures were made with cement to fly ash to sand proportions of 840:210:2100 when mixed for a target air content of 13% and water content adjusted for a flow of 100±5%. For mortars without fly ash or with fly ash that has a low admixture demand, the air content rises in approximate proportion to the amount of admixture used. For the other fly ashes there is virtually no increase in air content until some threshold dosage has been exceeded. Increments in dosages above that threshold result in curves essentially parallel to that of the mortar without fly ash.

These results show why in practical situations it may prove very difficult to control the air contents of concretes using fly ashes with high admixture

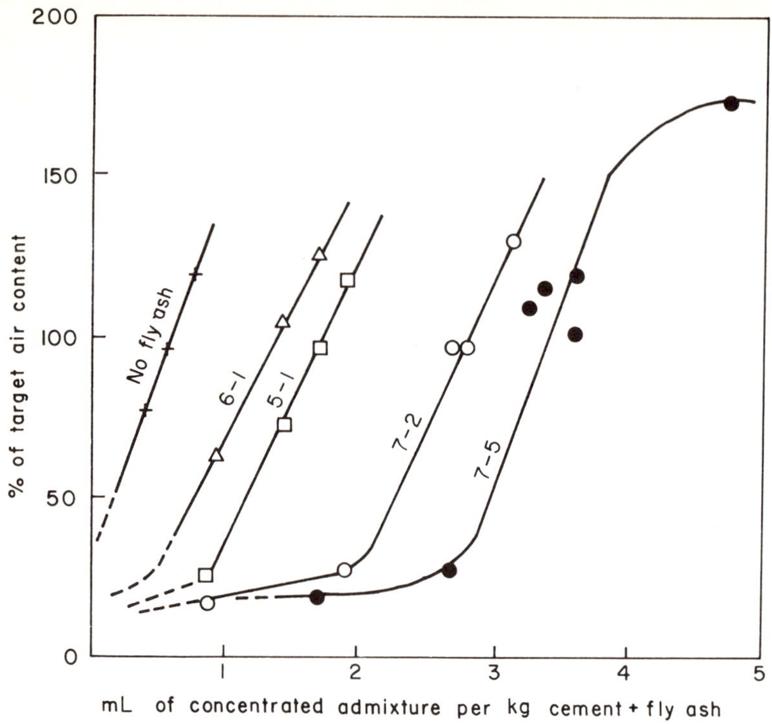

Fig. 5.11 Typical air content versus air-entraining agent dose rate curves for mortars with four different fly ashes. (Ref. 169)

requirements; if the air content of mortar made with fly ash No. 7-5 were 50% of the target and the dosage were doubled, the air content would be far in excess of the target. Trial and error would correct this problem, but if the fly ash also varies somewhat, adequate control may become very difficult. Fly ash No. 7-2 was a sample from the same source as No. 7-5 but from a different delivery.

Foam Index Test

In order to cope with such variability in AE admixture requirements of fly ashes, a simple test has been developed by Dodson[170] and slightly modified by Meininger.[169] In this test 50 mL water is added to 16 g cement and 4 g fly ash in a 4-oz (120 mL) glass bottle, which is capped and vigorously shaken for 1 minute. Diluted AE admixture is added in small increments, and the bottle is shaken vigorously for 15 seconds after each addition. The bottle is then laid on its side to see if the foam is stable; if the foam remains stable after 45 seconds, the amount of AE admixture used is the *foam index* of the fly-ash cement.

Results obtained by Meininger using the foam index test are compared with the AE admixture requirements for mortars made using 16 different samples of fly ashes in Fig. 5.12. Since the relationship between the foam index and the

Table 5.4 Fly Ash Properties[169]

Source ID No. (a)	Loss on ignition, %(b)	Organic, %(c)	Specific gravity (b)	Blaine fineness (b)		Passing No. 325, % (b)
				m^2/kg	$m^2/m^3 \times 10^{-3}$	
1-1	2.3	0.2	2.48	254	629	82
2-1	2.3	0.6	2.04	253	515	62
3-1	2.9	1.0	2.24	408	916	62
4-1	4.0	0.6	2.17	287	622	70
4-2	4.4	0.65	2.20	273	601	71
5-1	4.2	0.8	2.27	301	684	78
6-1	4.8	0.8	2.38	383	912	83
7-1	2.9	0.5	2.29	273	601	93
7-2	5.1	0.6	2.41	350	844	92
7-3	5.1	1.5	2.36	395	933	92
7-4	5.3	1.3	2.38	391	931	92
7-5	6.2	1.7	2.42	416	1008	91
8-1	4.7	0.5	2.40	386	927	87
9-1	5.9	1.0	2.24	305	682	69
10-1	13.7	1.8	2.12	456	966	69
10-2	16.9	2.6	2.14	510	1094	69

(a) Fly ashes from 10 sources; in some cases samples were obtained from more than one shipment from that source: two from Source No. 4, five from Source No. 7, and two from Source No. 10.
(b) Average of two tests.
(c) Organic matter determined by overnight soaking in solution of sodium dichromate and sulfuric acid, followed by titration using ferrous ammonium sulfate.

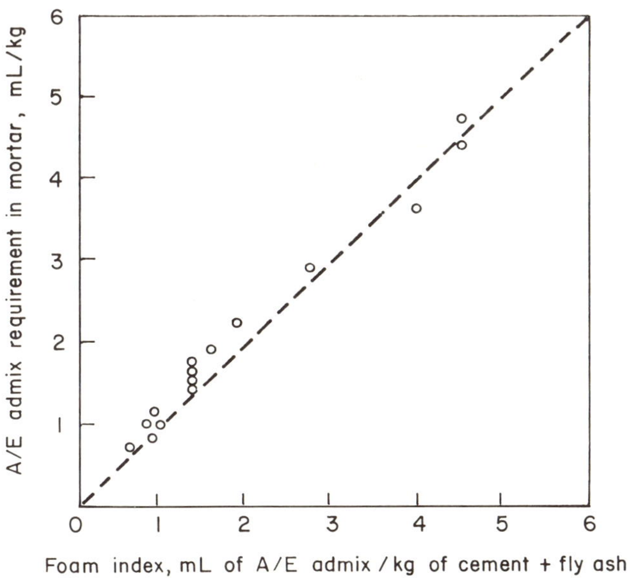

Fig. 5.12 Air-entraining agent requirement in mortars versus foam index at approximately 5 minutes. (Ref. 169)

admixture requirement in the mortar was very close to 1:1, it was concluded that the foam index test would be a good practical quality control tool.

Loss of Air

Meininger also examined the loss of air with time as another factor of importance in control of air content. He concluded that fly ash concrete is much more likely to have air loss problems than concrete without fly ash and that the air loss depends upon the properties of the fly ash, the AE admixture, and particular combinations of cement and sand. In mortars mixed for 1 hour with the amount of admixture required for 13% initial air contents, the mortars containing fly ashes with high admixture requirements also lost air more rapidly; many of them lost more than 30% of the initial air content. The losses were even greater after 2 hours.

It is widely believed that the carbon content of a fly ash is an indicator of its admixture requirement, because at least some of the carbon has very high specific surface area and sorption capacity.[12,55] Dodson[170] compared air content of mixes made with 14 different fly ashes with their loss on ignition at 750±50° C. Both Class F and Class C fly ashes were included, and loss on ignition values ranged from about 0.2 to 5%. The correlation was not very good, but a general trend indicated lower air contents with higher loss on ignition. Meininger, however, obtained reasonably good results by comparing the air loss data with the easily decomposed organic matter as determined by a sodium dichromate digestion test. Results for mortars made with two different sands and air-entraining admixtures with different fly ashes are shown in Fig. 5.13. This method appears to be a much better indicator of admixture adsorption capacity of the fly ash than the loss on ignition.

In Chapters 4 and 5 consideration of chemical reactivity of fly ashes and portland cements was deferred in order to concentrate on the physical structure and characteristics of their fresh mixtures. This structure is the reaction environment in fresh paste, mortar, and concrete. In the next chapter, the hydration chemistry of portland cements is briefly summarized prior to consideration of the pozzolanic reactions of fly ashes in Chapter 7.

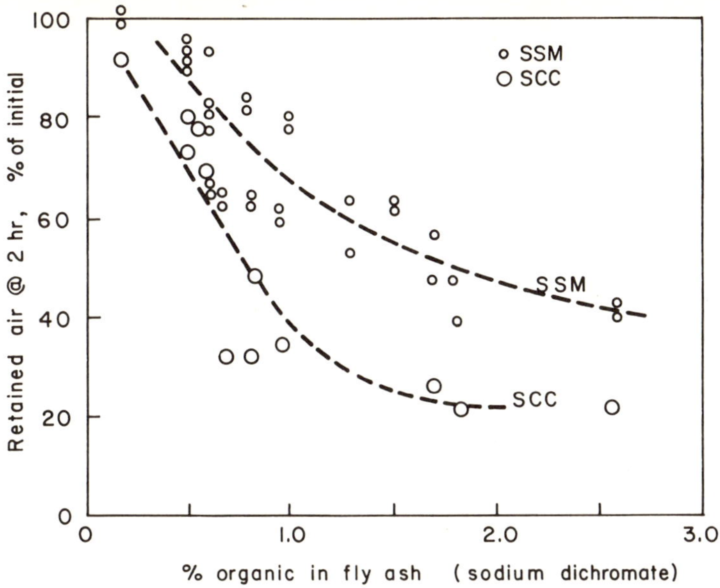

Fig. 5.13 Relationship between loss of air content and percent organic matter as determined by slow sodium dichromate reaction. SSM and SCC are mortars made with the same cement but with different sands and different air-entraining admixtures. (Ref. 169)

Chapter 6

Hydration of Portland Cement

Portland cement is a hydraulic cement; it sets, hardens, and does not disintegrate in water. Setting and hardening results from reaction of the compounds in the cement with water to form a dense structure of hydrated compounds.[171a] Portland cement is manufactured by making cement clinker from the raw materials in a cement kiln and, after cooling, intergrinding the clinker with a small percentage of some form of calcium sulfate, usually natural gypsum rock.[6] The clinker is composed of microscopic crystals of four major phases and several other phases, including calcium oxide and magnesium oxide, in minor amounts. Each phase contains dissolved minor constituents. The two most abundant phases are known as alite and belite, which are tricalcium and dicalcium silicates (C_3S and β-C_2S) containing minor amounts of magnesium, aluminum, iron, and other oxides. These crystals are embedded in a matrix of interstitial tricalcium aluminate (C_3A) and ferrite phases (e.g., C_4AF). Grinding the cement to a fine powder creates sufficient surface area for reactions to proceed at a rate adequate for most applications when the cement is mixed with water.

Hydration Chemistry

The hydration of portland cement and the products of cement hydration were succinctly described by Seligmann and Greening[171b] in 1968 in the following two paragraphs:

> The early hydration reactions can be considered to be an essentially continuous sequence of phase equilibrium situations. The contact between cement and water immediately produces a set of reactions determined by local solution concentrations at the cement-water interfaces to produce materials such as calcium sulfoaluminate hydrates and a hydrogarnet. These local reactions are followed by a period of slow hydration in which the solution composition is relatively constant as ettringite precipitates and lime supersaturation persists. When the calcium hydroxide nucleates, a calcium silicate hydrate gel containing sulfate and a little alumina forms. Ettringite continues to form until the gypsum is depleted; the solution sulfate concentration drops. Rapid solution of the ettringite and the remaining aluminate and ferrite phases causes formation of a gel, richer in alumina, the calcium sulfoaluminoferrite hydrate, and the low (mono) sulfate form of calcium sulfoaluminate hydrate. The formation of further aluminasilica hydrogarnet at this stage may be due to local depletions of sulfate in the liquid phase. After the ettringite is depleted, the lowest sulfate form of calcium sulfohydroxyaluminate is ultimately produced.
>
> The solid phases at complete hydration in the cement-water system are identified as a calcium silicate hydrate gel containing appreciable quantities of alumina, sulfate, and alkalies, the low sulfate form of calcium sulfoaluminoferrite hydrate, the lowest sulfate form of calcium sulfohydroxyaluminate hydrate, an alumina-silica hydrogarnet, and calcium hydroxide. Ettringite or its iron analogue is also present in cements low in C_3A, but more often forms as a result of carbonation.

Research continues to provide more detailed information. For example, the concentration of calcium ions in solution does increase during the "period of slow hydration" (dormant period), and "calcium hydroxide nucleates" means stable nuclei begin to grow at significant rates. The effects of local departures from equilibrium are discussed later in this and other chapters. A recent review of cement hydration has been given by Skalny and Young.[172]

The most important reactions may be represented (in cement chemists' notation: C = CaO, S = SiO_2, H = H_2O, A = Al_2O_3, F = Fe_2O_3, $\overline{S}$ = SO_3) by the following simplified equations:[59,171,173]

$$2C_3S + 6H \rightarrow C_3S_2H_3 + 3CH \qquad (6.1)$$

(calcium silicate hydrate + calcium hydroxide)

$$2C_2S + 4H \rightarrow C_3S_2H_3 + CH \qquad (6.2)$$

$$C_3A + 3C\overline{S}H_2 + 26H \rightarrow C_3A(C\overline{S})_3H_{32} \qquad (6.3)$$

(gypsum) (ettringite)

$$C_3A + C\overline{S}H_2 + 10H \rightarrow C_3AC\overline{S}H_{12} \qquad (6.4)$$

(calcium monosulfoaluminate hydrate)

$$C_3A + CH + 18H \rightarrow C_4AH_{19} \qquad (6.5)$$

(calcium aluminate hydrate)

$$C_4AF + 2CH + 10H \rightarrow C_6AFH_{12} \qquad (6.6)$$

(hydrogarnet)

Calcium and magnesium oxide also hydrate to the hydroxides.

Equations (6.5) and (6.6) are examples of reactions in the absence of sufficient sulfate to form monosulfoaluminate or ettringite (or their iron-containing analogs) as in equations (6.3) and (6.4); the C_3AH_6 hydrogarnet forms at higher temperatures. It is important to note that reactions (6.5) and (6.6) consume calcium hydroxide directly and, therefore, compete with pozzolanic reactions for the lime released by hydration of the silicates. In addition, reactions (6.3) and (6.4) also require calcium and are affected by any process which lowers the calcium ion concentration.

The hydrate phase compositions are more complex than indicated in the simplified reaction equations above. The calcium sulfoaluminate and sulfoaluminoferrite hydrates are actually hydrates of variable composition denoted as *AFm* (analogous to monosulfoaluminate) and *AFt* (analogous to the trisulfate as in ettringite).

The calcium silicate hydrate composition is given only approximately by $C_3S_2H_3$. The calculated average molar lime-silica ratio varies from about 1.4 to 1.7 in pure calcium silicate pastes.[174] Because the actual compositions are indefinite, the common symbol used for the calcium silicate hydrate is C-S-H rather than a stoichiometric formula. Copeland, et al., also showed that the calcium silicate hydrates react with aluminates, ferrites, and sulfates to form substituted calcium silicate hydrate gels containing small amounts of the oxides of aluminum, iron, and sulfur[175] and that the C-S-H in cement pastes was likely to have a higher Ca to Si ratio than that of C_3S pastes.[176]

During the early hydration period, the composition of the liquid phase is controlled mainly by the solution of calcium, sulfate, sodium, and potassium ions. Very little alumina, silica, or iron are present in solution. Gypsum and calcium hydroxide separately have solubilities of 15 and 20 mmol/L at 25° C. The solution is supersaturated with respect to calcium hydroxide, especially before the nucleation of CH at the onset of setting. Alkalies released by the readily soluble alkali sulfate phases in cements affect these concentrations so that the composition tends to be governed by the equilibrium:

$$\underset{\text{(solid)}}{CaSO_4 \cdot 2H_2O} + \underset{\text{(aqueous)}}{2MOH} \rightleftharpoons \underset{\text{(aqueous)}}{M_2SO_4} + 2H_2O + \underset{\text{(solid)}}{Ca(OH)_2} \qquad (6.7)$$

in which M represents the alkalies. At advanced stages of hydration of low water-cement ratio pastes, the alkali concentration may exceed 0.5 N with a pH above 13, even with a low alkali cement. At such concentrations appreciable quantities of alkalies can be incorporated into the C-S-H.

The effect of alkalies and sulfates on the hydration of cements has been reviewed.[172,177] Alkalies can be present in cement clinkers in the silicate phase as alkali belite ($KC_{23}S_{12}$), in the aluminate phase as alkali aluminate (NC_8A_3), or as alkali sulfates. If there is sufficient sulfate in the clinker, most of the alkalies will be present as readily soluble sulfate phases such as potassium sulfate ($K\overline{S}$), aphthitalite ($N_xK_y\overline{S}$), or calcium langbeinite ($2C\overline{S}\cdot K\overline{S}$). Alkalies in the silicate phases will be released during hydration much more slowly.

Alkali sulfates accelerate the hydration of cement, affecting both the alite (C_3S) and C_3A hydration rates. There are several interactions in these reactions, but the main effects may be explained in simplified terms as follows. Skalny and Young[172] suggest that the acceleration of the alite is a result of the influence of alkali salts on calcium hydroxide solubility. Alkalies decrease the concentration of calcium in solution (Fig. 7.4) and increase the silicate ion concentration. The reduction of the calcium concentration should also accelerate the C_3A hydration since an adequate supply of both calcium and sulfate in solution is essential for control of the early C_3A reaction by formation of an ettringite coating, as in Eq. 6.3. Potassium sulfate can also react further with gypsum to form syngenite ($K\overline{S}\cdot C\overline{S}\cdot H$).

$$K\overline{S} + C\overline{S}H_2 \rightarrow K\overline{S}\cdot C\overline{S}\cdot H + H \qquad (6.8)$$

For best strength development and volume stability, the amount of gypsum added is increased with the aluminate content of the cement. There is an optimum gypsum content for each cement, which is usually at a sulfate to aluminate molar ratio of about 0.6 for cements of ordinary fineness. Although alkalies accelerate the reactions and strength development at early ages, they usually reduce strengths at 28 days and beyond. Skalny and Young suggest this reduction may be associated with changes in optimum gypsum requirements with alkali content, which in turn are related to cement fineness. They concluded from a recent study of earlier work by Lerch[178a] that fineness is the major parameter controlling optimum gypsum content. Because of the tendency for modern clinkers to contain larger quantities of alkali sulfates, gypsum requirements were recently reviewed by Kanare and Gartner.[178b] Because fly ashes contribute variable amounts of alkalies and sulfates, such studies are also important with respect to the portland cement-fly ash hydration interactions.

Actual Phase Compositions

Work by Lachowski, Mohan, and Taylor[179] on hydrated ground clinker and portland cement pastes has provided quantitative compositional data on individual solid particles from which mean values and ranges of composition were determined. Hydration of the normally ground portland cement produced particles of C-S-H, CH, AFm, and AFt which could usually be characterized by morphology and electron diffraction. Of 408 particles analyzed, only 43 could not be so characterized; they were probably composites of the others and unhydrated clinker.

The Ca to Si atom ratios of the C-S-H particles in 28-day cured cement pastes ranged from 1.27 to 2.11. The mean Ca to Si ratios for different pastes ranged from 1.6 to 2.4 with an overall mean of 2.0 for 156 analyses. The median Al to Si, Fe to Si, and S to Si ratios in the C-S-H were 0.09, 0.02, and 0.06, respectively, for the 28-day-cured cement pastes. Nearly all the C-S-H particles contained smaller amounts of Mg. The Fe to (Al + Fe) ratios of the unhydrated cements affected the individual ratios of Al and Fe relative to Ca + Mg, but not their sum, suggesting that the Al and Fe compete for similar sites in the C-S-H gel structure. The S to (Ca + Mg) mean ratios decreased from 0.11 to 0.05 between 7 and 28 days of curing. The mean ratio of (Si + Al + Fe + S) to (Ca + Mg) in all the pastes was 0.66, very nearly the same as the mean Si to Ca ratio of the C-S-H in C_3S pastes, 0.65. This similarity supports the idea that the SO_4^{2-} ions and Al- and Fe-containing complexes all replace silicate in the C-S-H gel. Calcium hydroxide constitutes about 20% of the hydration product of ordinary portland cements. Analysis showed no Mg in the calcium hydroxide particles, but they contained measurable amounts of Si, Al, S, and Fe. The mean Ca to Si ratio was 32, with lesser amounts of Al, S, and Fe.

Analyses of the AFm phase particles showed that they were not pure monosulfoaluminate hydrate but contained almost as much Si as S and much more Al than expected for monosulfoaluminate. The AFm phases have the general formula $Ca_2R(OH)_6X_x$ aq., in which R is an octahedral cation (Al^{3+} or Fe^{3+}) and X an interlayer anion ($(OH)^-$, SO_4^{2-}, $Al(OH)_4^-$, $H_2AlSiO_5^-$). The latter anion results in C_2ASH_8, known as gehlenite hydrate, a product often identified in fly ash-cement pastes. The AFm phase can be regarded as a mixture or solid solution of C_4AH_x, $C_4A\bar{S}H_x$, and a composite component containing Al and Si.

The AFt phases have the general formula $Ca_3R(OH)_6Y_x$ aq., where *Y* is an anion in a channel site which can be occupied by SO_4^{2-}, a silicate, water molecules, and possibly OH^-. The mean ratios of S to 2(Al + Fe + Si) are always below that in ettringite, 3.0. This ratio decreases with curing time after 1 day.

Lachowski et al. concluded that differences in composition were a result of the difficulty of ionic transport within the pastes and that equilibrium is attained only locally on a scale of microns or less. Nevertheless, comparisons of the phase compositions indicated that nearly all of the S and about one-third of the Al end up in the C-S-H. The Fe also tends to enter the C-S-H in preference to the AFm or AFt phases.

Structure and Porosity of Hardened Cements

The hydration and pozzolanic reactions produce hard solids from soft plastic pastes because the reaction products form bridges of solid hydrates across the water-filled spaces between the cement, pozzolan, and aggregate particles. These space-filling processes can be described in terms of the morphology and distribution of the solid hydrates or in terms of the progressive filling of the pore spaces. The latter approach has long been used to explain the properties of hardened portland cement pastes, because the microstructure of the C-S-H gel is on a scale that was not suited to available methods of observation and measurement. The results presented in the last section show that modern methods like TEM and SEM now provide much information about the phase compositions during the structure formation processes. In a review of the microstructure of hardened cements, Diamond[(180)] proposed a classification system to describe four main morphologies observed in SEM studies of pure calcium silicate and portland cement pastes:

Type I	Fibrous particles
Type II	Reticular network (interlocking or honeycomb structure)
Type III	Small, irregular flattened grains, typically 0.3 μm across
Type IV	Inner product, C-S-H formed within the original particle boundary with a dimpled appearance due to packed small grains or regular pores about 0.1 μm in size

In addition, the existence of "bulk" C-S-H developing at later stages of hydration, with no discernible particle morphology, was also recognized. Unfortunately, these methods have not yet advanced to the stage whereby the structure of the "bulk" C-S-H hardened gel can be directly observed and the strength or other properties predicted or quantitatively correlated with the observations. For such relationships porosity measurements are able to provide quantitative data. Powers and his co-workers were among the first to show how mechanical strength and other physical properties, such as permeability, depended upon changes of water-cement ratio and degree of hydration by calculating the effect of these changes on the extent of filling of the pore space with hydrates.[(181)]

Powers' model may also provide a reasonable first approximation for studies of portland-pozzolanic cements. The basic idea in Powers' model is that cement hydration gradually fills the originally water-filled space with hydration products in the form of a cement gel, a material of characteristic minimum porosity and high specific surface area. The minimum porosity of the cement gel was determined to be about 28%. If the original water-cement ratio is less than about 0.38, the paste will hydrate until the original pore space is filled with gel. At that time the paste will have a porosity lower than the characteristic porosity (0.28) of the gel because of the presence of the unhydrated cement remaining. The minimum gel porosity corresponds to an amount of pore water equal to 3.0 V_m, where V_m is the mass of a monolayer of water adsorbed on the surfaces of the pores. The specific surface area is about 200 m^2/g after the paste has been compeletely dried. The pores in the gel were estimated to be so small (less than about 3 nanometers) that most of the water contained in them is strongly affected by the solid surface force fields.

Surface area measurements by small angle X-ray scattering have shown that, before drying, the specific surface area is about 700 m^2/g.(182a,b) It has been suggested that the change of structure during drying is the cause of the irreversible component of drying shrinkage.(183a,b)

The porosity in excess of the gel porosity, i.e., the porosity resulting from the pore space not filled by the gel, is termed capillary porosity. Thus, the evaporable water, w_e, in the paste can be divided into the gel water, w_g, and the capillary water, w_c. Capillary pores were found to be generally much larger (about 0.1 μm and larger) than the gel pores and interconnected by smaller capillaries and gel pores.(184) However, there is no sharp size distinction between capillary and gel pores; the densest cement gels always contain some of the larger pores as well. By Powers' definition, these large pores in dense gels are considered part of the characteristic porosity of the gel.

The importance of Power's model is that it permits the development of quantitative relationships between the original water-cement ratio and the degree of hydration of the cement to properties of engineering interest such as compressive strength, permeability, and freezable water content. In Chapter 8 we shall review work on calculation and measurements of gel and capillary porosities in portland and blended cements and how these have been related to strength. However, much of the research cited above and in Chapter 8 has been done with cement pastes cured so as to remain fully water saturated during hydration. If the paste is sealed to prevent drying, but cured without access to additional water, the pores are partially emptied by hydration because the hydrates occupy less space than the cement plus water before reaction. This is called self-desiccation.

If partial drying by loss of moisture also occurs, even more of the pore space is emptied. Such drying processes can strongly affect the resulting pore structure and properties because hydrates form only in the water-filled portions of the pore structure. Space filling of the largest pores will be most affected by drying because they are the first to be emptied during drying. The importance of moist-curing of concretes is that the physical properties and durability depend strongly on the extent to which the large pores become filled with hydrates. In the case of concretes in which fly ash has been partially substituted for portland cement, curing is especially important because of the reduced cement content and slowness of the pozzolanic reactions.

Chapter 7

Pozzolanic Reaction Chemistry

Definitions

The term "pozzolan" is employed to designate a siliceous or a siliceous and aluminous material which in itself possesses little or no cementitious value but will, in finely divided form and in the presence of moisture, chemically react with calcium hydroxide at ordinary temperatures to form compounds possessing cementitious properties.[1] Such materials are said to have "pozzolanic activity."

By the ASTM definition above, some fly ashes (Class F) are pozzolans, whereas some (Class C) have cementitious properties as well. Fly ash is sometimes regarded as an "artificial pozzolan," but Sersale[127] has objected to this usage because it tends to create confusion with materials of distinctly different origins, microstructures, and properties. Fly ash is his preferred term. There is also confusion with respect to the use of the term "pozzolanic reactivity." Takemoto and Uchikawa[185] use the term to indicate the degree of the chemical reaction of the pozzolan and water with calcium hydroxide, or material which produces calcium hydroxide in water. Clifton et al.[186] use the term for the rate of reaction to produce strength-giving phases. The ASTM standard "pozzolanic activity indexes," which are determined from strength tests, are measures of reactivity with calcium hydroxide or given portland cements. ISO uses the term "pozzolanicity" for the results obtained in tests to determine the reduction of the concentration of calcium dissolved in the liquid of pozzolanic cement suspensions.[187]

Calcium Silicate Hydrates at Ordinary Temperatures

The major pozzolanic reaction products are substituted calcium silicate hydrate gels with additional anions and cations incorporated into the structure. Since these are systems of considerable complexity, it is useful to consider first some closely related simple hydrates[188] which are also identified in lime-pozzolan or cement systems hydrated at ordinary temperatures. Elevated temperatures promote the formation of different, well-crystallized products.

The reaction of calcium hydroxide with silica gel produces two calcium silicate hydrates known as C-S-H I and C-S-H II. C-S-H I consists of poorly crystallized material of characteristic crumpled foil morphology as seen in the electron microscope. The CaO to SiO_2 molar ratio ranges from 0.8 to 1.5. C-S-H I may be produced in a variety of other ways, such as by repeated extraction of C_3S or β-C_2S with water. A representation of a type of disordered structure suggested for C-S-H I is shown in Fig. 7.1.[189]

C-S-H II is a semicrystalline material of a characteristic fibrous structure and a molar CaO to SiO_2 ratio of 1.5 to 2.0. Both C-S-H I and C-S-H II have X-ray diffraction (XRD) patterns in which most of the spacings approximate those in

Fig. 7.1 C-S-H I: diagrammatic sketch showing possible division of a single tobermoritelike sheet into domains characterized by angular displacement of the chain direction through about 60°. Dotted areas represent very disordered regions. (Ref. 189)

the natural mineral tobermorite. C-S-H II also forms from C_3S or β-C_2S hydrating in excess water at room temperature.

C-S-H gel is the major product of paste hydration of C_3S, β-C_2S, or portland cement at room temperature. It is very poorly crystallized, showing no distinct structure under the microscope; only a mean refractive index (1.5 to 1.55) can be obtained. The XRD pattern shows no more than three weak and diffuse peaks at about 0.305, 0.28, and 0.180 nm, close to spacings in C-S-H I and C-S-H II. Basal spacings are not usually detected for C-S-H gel but may appear if the contact solution contains appreciable alkali.[190] The molar CaO to SiO_2 ratios are variable and depend upon the composition of the contact solution.

Phase Equilibria

The relationship between the calcium silicate hydrate composition and the concentration of CaO in solution in equilibrium at room temperature is shown in Fig. 7.2. The vertical lower portions of the curves indicates that there is an invariant point between hydrated silica and a monocalcium silicate hydrate at about 1-2 mmol/L (0.06-0.11 g CaO/L). Below this lime concentration, the concentration of silica can rise to about 0.3 g SiO_2/L, but at higher lime concentrations the silica concentrations are very low (below 0.006 g SiO_2/L), and the CaO to SiO_2 ratio of the solid is near 1.0 or above. Over a range of lime concentrations from 0.05 g CaO/L up to near saturation, C-S-H I is formed with CaO to SiO_2 ratios ranging from 0.8 to 1.5, while in the region of lime saturation C-S-H II is formed with CaO to SiO_2 ratios of 1.5 to 2.0; their distinction is almost impossible to detect at $C_{1.5}SH_x$. The sharp rise in the lime-silica ratio at lime solution concentrations near saturation (20 mmol CaO/L) is evidently a result of adsorption of lime.

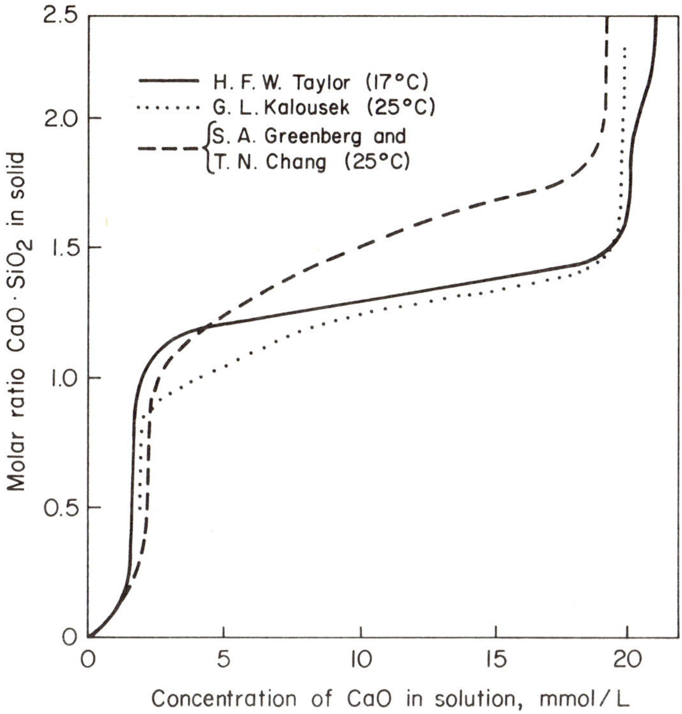

Fig. 7.2 Composition of hydrous calcium silicates in equilibrium with lime solutions at various concentrations. (Ref. 191)

It has been suggested[189] that in this range of compositions the hydrates are an intimate mixture of C-S-H I and C-S-H II. The fibrous C-S-H II could then be thought to form on substrates such as those indicated in Fig. 7.1. Fig. 7.3 is an interesting SEM micrograph[192] which suggests such a structure in which fibrous material is arranged at preferential 60° angles. It may also be that the original structure was dehydrated and altered by the electron beam and high vacuum in the SEM. Perhaps the C-S-H I sheets shrank and retracted onto reinforcing fibers of C-S-H II.

Greenberg and Chang[193] determined the average solubility product for the CaO to SiO_2 range from 0.2 to 1.8 to be

$$K_{sp} = a_c \times a_s = 10^{-7.2} \tag{7.1}$$

in which a_c is the activity of Ca^{++} and a_s is the activity of $H_2SiO_4^{2-}$. Dron[194] has shown that, in the pH range typical of the pure C-A-S-H system, it is justifiable to neglect the $H_2SiO_4^{2-}$ with respect to $H_3SiO_4^-$; the solubility product for C-S-H becomes $10^{-8.3}$ if expressed in terms of this ion.

In a classic study of the Na_2O-CaO-SiO_2-H_2O system, Kalousek[195] prepared alkali-calcium-silicate hydrates by reaction of calcium hydroxide with sodium

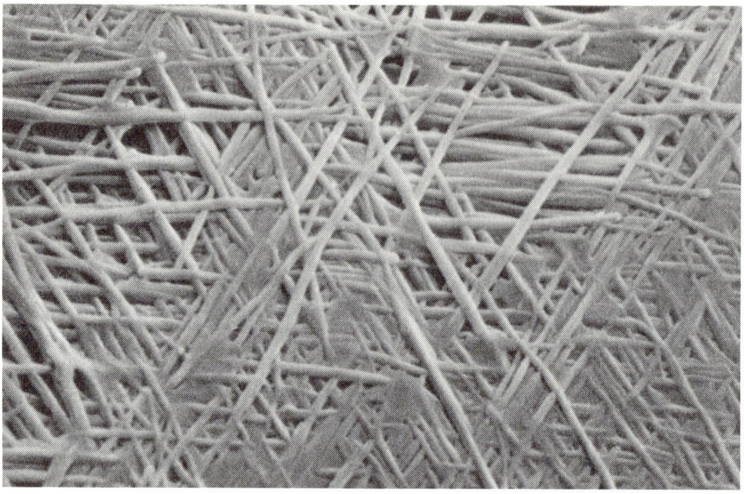

Fig. 7.3 SEM micrograph at 20,000X of hardened cement paste prepared using a blast furnace slag cement. (Ref. 192)

silicate solutions. In this way equilibrium could be established within weeks. Slow changes in crystallinity over long periods, however, could produce very different equilibrium points. Kalousek determined the effect of sodium concentrations on the calcium and silica concentrations in solutions in contact with the solid phases; he also determined C to S and N to S (Na_2O to SiO_2) ratios of the solid phases as shown in Fig. 7.4. The curves show that, as the alkali concentration increased from 0 to 15 g Na_2O/L (0.5 N NaOH), the calcium in solution decreased to low levels (common ion effect), while the N to S ratios of the C-S-H increased to about 0.2 mole/mole and the C to S ratio decreased from about 2.0 to about 1.0. At higher alkali concentrations, the silica in solution increased substantially while the other quantities changed very little.

Dron[194] constructed the equilibrium phase diagram for the $CaO\text{-}Al_2O_3\text{-}SiO_2\text{-}H_2O$ system at ordinary temperatures from solubility products for C_2AH_8, C_4AH_{13}, CS_2H, CSH (C-S-H in our notation), CH, AH_3, SH, and C_2ASH_8, which are taken to be the thermodynamically stable compounds. Prior work had established all of the solubility products except that for gehlenite hydrate, C_2ASH_8, which was then determined experimentally. The resulting set of equations was solved in terms of the concentrations of the dissolved species:

$$\begin{aligned} x &= [Al(OH)_4^-] \\ y &= [SiO(OH)_3^-] \\ z &= [Ca^{2+}] \end{aligned} \tag{7.2}$$

These concentrations define surfaces for each equilibrium phase. Only three variables are necessary, because the OH^- concentration is obtained from the condition for electric neutrality. The intersections of the surfaces define the monovariant lines of coexistence of two solid phases in equilibrium with the

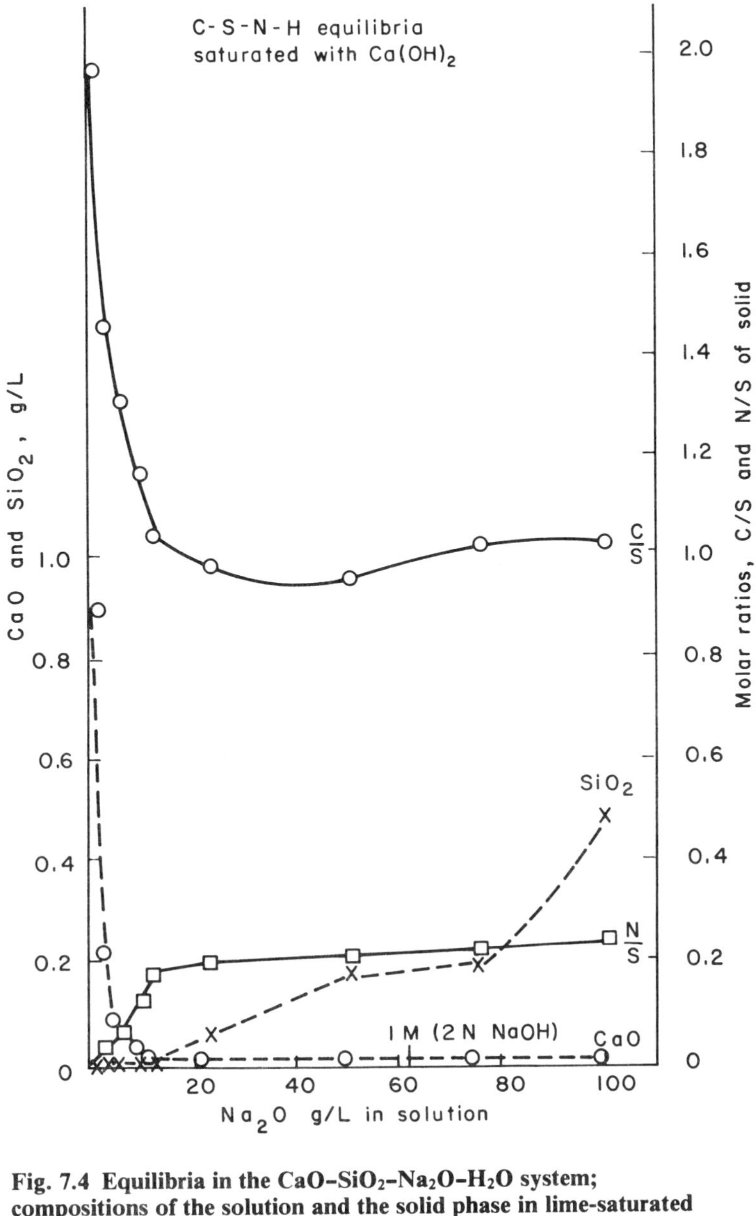

Fig. 7.4 Equilibria in the $CaO-SiO_2-Na_2O-H_2O$ system; compositions of the solution and the solid phase in lime-saturated liquid phase at various Na_2O concentrations. (Ref. 195)

solution of the given composition. The monovariant lines converge in groups of three at the invariant points of the system where three solid phases coexist with the solution. The surfaces in this way define the three-dimensional phase diagram represented in perspective in Fig. 7.5. The diagram clearly shows that for calcium hydroxide to form or remain under equilibrium conditions, the Al and

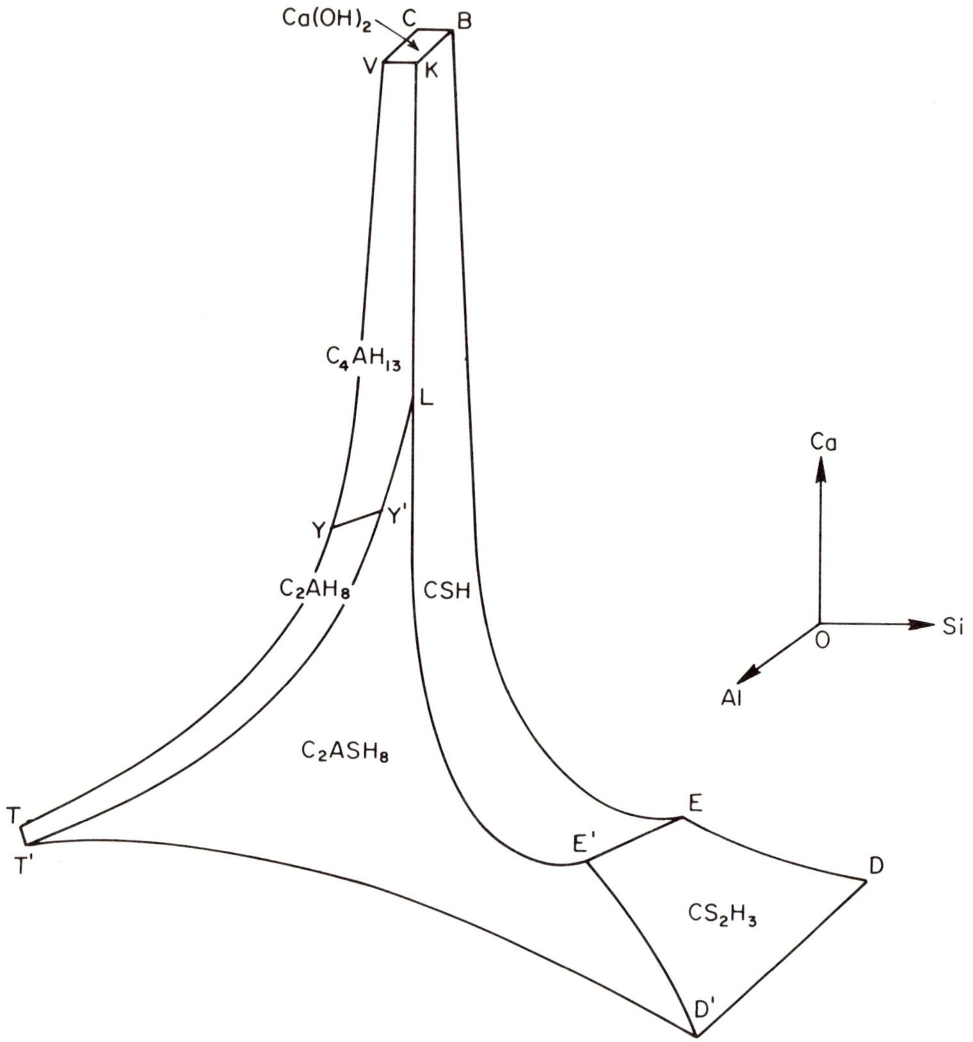

Diagram uses both cement chemists' and conventional notation.

Fig. 7.5 Phase equilibrium diagram for the CaO–Al_2O_3–SiO_2–H_2O system. (Ref. 196)

Si concentrations in solution must be very low and the Ca concentration relatively high. At higher Al or Si concentrations, C_4AH_{13} or C-S-H will form and CH will disappear, as the Ca concentration falls. At still higher Al and Si concentrations, other phases will form.

The figure shows six invariant (triple) points; of most practical interest are the points K and L. Point K represents the coexistence of CH, C_4AH_{13}, and C-S-H; point L represents the coexistence of C_4AH_{13}, C-S-H, and C_2ASH_8.

In order to generalize the diagram to systems including other ions, such as Na^+ and SO_4^{2-}, Dron adopted another coordinate system:

$$x = \log \frac{[Al(OH)_4^-]}{[OH^-]}$$

$$y = \log \frac{[SiO(OH)_3^-]}{[OH^-]} \qquad (7.3)$$

$$z = \log [Ca^{2+}][OH^-]^2$$

The projection of the resulting three-dimensional diagram on the *xy* plane yields Fig. 7.6. Both of these diagrams clearly show the incompatibility between calcium hydroxide and gehlenite hydrate, which cannot coexist at equilibrium. C-S-H, on the other hand, has the largest field of stability and can coexist with CH, C_4AH_{13}, C_2ASH_8, and CS_2H. It should also be noted that the hydrogarnet phases (such as C_3AH_6) are not shown explicitly but should exist at slightly elevated temperatures and low sulfate concentrations.

The presence of sulfates leads to the formation of additional calcium sulfoaluminate hydrate phases, such as the monosulfoaluminate and ettringite (see "Hydration Chemistry" in Chapter 6 and "Pozzolanic Reactions" in Chapter 7). Ettringite, however, is known to be unstable in the presence of high silica pozzolans in water, even in water saturated with respect to gypsum, but decomposition is largely prevented in lime-saturated water.(197)

Low concentrations of alkali in solution do not change the general nature of the equilibrium relationship between the calcium aluminate and sulfoaluminate solid phases, but they substantially alter the liquid compositions because of the increased solubility of SiO_2 and Al_2O_3 and the decreased solubility of CaO.(198)

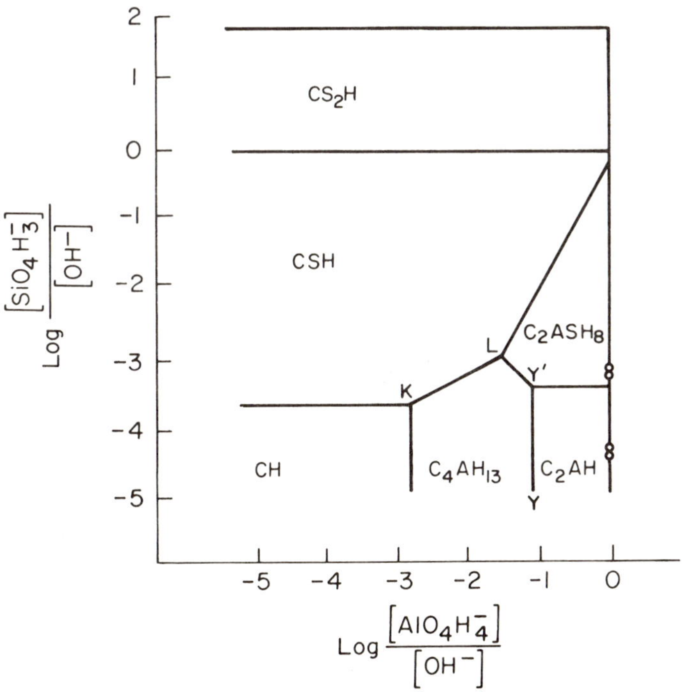

Fig. 7.6 Projection of a generalized $CaO\text{-}Al_2O_3\text{-}SiO_2\text{-}H_2O$ phase equilibrium diagram. (Ref. 194)

Pozzolanic Reactions

The most important pozzolanic reactions are those between reactive silica and calcium hydroxide that produce calcium silicate hydrates.[127,197,199–201] For pure materials, the reaction can be represented by the simplified equation

$$xCH + yS + zH \rightarrow C_xS_yH_{x+z} \quad \text{(calcium silicate hydrate)} \qquad (7.4)$$

Studies of many lime-pozzolan systems have indicated the hydrate to be C-S-H I.[127,201] However, recent studies[202] have shown that before reaching equilibrium the lime-silica ratio of x to y ranges from 0 to 2.0 depending on the local concentration of reactants in the mixture, as will be shown in "Hydration and Pozzolanic Reactions" in Chapter 8.

In addition, the alumina in the pozzolan may yield a variety of hydrates. In one recent study on the characterization of the pozzolanicity of fly ash, aluminum hydroxide was used as the reference pozzolanic material.[203] Reaction products include calcium aluminate hydrate (C_4AH_{19}), gehlenite hydrate (C_2ASH_8), ettringite ($C_3A{\cdot}3C\bar{S}{\cdot}H_{32}$), and calcium monosulfoaluminate hydrate ($C_3A{\cdot}C\bar{S}{\cdot}H_{12}$). If gypsum is present, the monosulfoaluminate reaction may be represented in this way:

$$A + 3CH + C\bar{S}H_2 + 7H \rightarrow C_3AC\bar{S}H_{12} \qquad (7.5)$$

with analogous reactions for ettringite and calcium aluminate hydrate.

Calcium carboaluminate hydrates are always possible reaction products because of reactions with atmospheric carbon dioxide and limestone mineral admixtures or aggregates. Carbonation should tend to stabilize ettringite with respect to transformation to the monosulfate.[171b]

Burned kaolinite ($Al_2O_3{\cdot}2SiO_2$), in itself a useful artificial pozzolan, is not very different in composition from some high alumina fly ashes (54% SiO_2, 46% Al_2O_3). For that reason, "synthetic fly ashes" are sometimes prepared from kaolinite for chemical studies.[123,204] The reaction with calcium hydroxide may produce both C-S-H and gehlenite hydrate[205] and, when the calcium hydroxide is completely reacted, may be represented by

$$AS_2 + 3CH + zH \rightarrow CSH_{z-5} + C_2ASH_8 \qquad (7.6)$$

Sersale and Orsini[206] concluded that the main compounds formed both in suspensions and in lime-pozzolan pastes are C-S-H and C_2ASH_8, although C_4AH_x and the solid solution with end members C_3AS_3 – C_3AH_6 (garnet-hydrogarnet) are formed under certain conditions. High alumina glasses tend to form gehlenite hydrate if silica is correspondingly high or to form hydrogarnets even at room temperature if the lime content of the system is high. In the presence of gypsum, ettringite and monosulfate form instead of the gehlenite hydrate. Fly ash containing 1.39% SO_3 also yields ettringite.[205] The ultimate reaction products depend on the proportions of lime and pozzolans present, a matter which will be discussed in Chapter 8 (see Fig. 8.9).

Minnick observed that lime (and magnesium oxide) can react with iron oxide in the glassy phase of fly ash.[207] However, Watt and Thorne showed that fly ashes that reacted with lime showed a progressive increase in the amounts of reacted silica and alumina but little increase in the amount of reacted ferric oxide.[123] Both observations may be correct, because there is little iron in the glassy phase and, as we shall see, iron in the glassy phase decreases its reactivity.

Energetics

Kondo and Ohsawa[208] tabulated the enthalpy and Gibbs free energy changes for 19 different pozzolanic reactions. The values of ΔH_{298} for the reactions of silica glass and quartz with $Ca(OH)_2$ are given as –3.4 and –2.3 kcal/mole $Ca(OH)_2$, respectively. The corresponding values for the hydration of CaO, C_3S, and β-C_2S are about –15.5 kcal/mole CaO, –21.4 kcal/mole C_3S, and –5.2 kcal/mole C_2S,[209] showing that the additional heat liberation associated with the pozzolanic reaction is much less than that of the cement hydration reactions. The values of the Gibbs free energy changes show that the driving force for the reaction of calcium hydroxide with silica glass [$\Delta G_{298} = -4.2$ kcal/mole $Ca(OH)_2$] and with quartz (–3.3 kcal/mole) are both rather low but greater than for the hydration of β-C_2S. The ΔG values for the hydration of C_3S and β-C_2S have been estimated to be –16 kcal/mole C_3S and -2.2 kcal/mole C_2S.[209] Values given for ΔH and ΔG for the reaction of $Al_2O_3 \cdot 3H_2O$ and $Al(OH)_3$ (gel) with calcium hydroxide were either very small negative values or else positive, indicating that the reactions would not proceed spontaneously. This finding may be inaccurate because the values were calculated as small differences between large numbers.

Lime-Pozzolan Reaction Mechanisms

Pozzolanic reactions involve chemical mechanisms similar to the corrosion of glass and alkali-aggregate reactions of reactive siliceous aggregates in concrete. In silica glasses, the silica tetrahedra share oxygen atoms in a three-dimensional network. Each of the four O^{2-} ions is linked to the central Si^{4+} in two adjacent tetrahedra. However, on the glass surfaces there are unsatisfied oxygen and silicon bonds which take up water to form OH^- ions. Sersale has summarized one suggested mechanism for such reactions in which other O^{2-} ions in the glass in contact with water also transform progressively to OH^- ions.[127] The surface tetrahedra are thus less and less well bonded until they break away and pass into solution as $H_3SiO_4^-$. If the solution contains Ca^{2+} ions, calcium silicate hydrates are formed.

Takemoto and Uchikawa[185] summarized the mechanism of the paste hydration of pozzolans with calcium hydroxide. After the pozzolan is mixed with water, the liquid quickly becomes saturated with respect to solid calcium hydroxide so that the pH is kept above 12.7. In this high pH solution the SiOH groups on the silica surface are dissociated to SiO^- and H^+, leaving negatively-charged grains. Ca^{2+} ions are adsorbed on the negative surface and alkalies in the glass dissolve into the liquid phase. There remains a thin amorphous Si and Al-rich layer on the surface. SiO_4^{4-} and AlO_2^- in the layer gradually dissolve and combine with Ca^{2+}, forming solid hydrates that increase the thickness of the layer. This layer is repeatedly swelled and destroyed by osmotic pressure so that the reaction continues. Because the alumina diffuses more rapidly than the silica and generally requires a higher calcium ion concentration for the formation of hydrates than do the calcium silicate hydrates, the calcium silicate hydrates tend to form on the glass while the aluminate hydrates form at some distance.

If fly ashes were simple homogeneous materials, it would be logical at this point to discuss the kinetics of the chemical reactions. Clifton et al.[(186)] loosely defined the reactivity of a fly ash as the rate at which strength-giving phases are produced by the reactions between the fly ash and a portland cement in the presence of water. However, the rates of the reactions depend on many factors other than the composition and intrinsic characteristics of the fly ashes. The results obtained depend strongly on the reaction conditions and methods of measurement, as will be seen in Chapters 8, 9, and 10.

Chapter 8

Fly Ash-Cement Hydration Interactions

Hydration and Pozzolanic Reactions

The cement hydration and pozzolanic reactions in fly ash-portland cement-water mixtures do not proceed independently. The cement hydration rate is accelerated by the presence of fly ash during most of the reaction but may be initially retarded. Readily soluble calcium sulfate may provide more rapid establishment of control of the initial rapid hydration of the C_3A. Alternatively, soluble alkalies on the fly ash particles may accelerate the reaction. Soluble organics may be adsorbed, interfering with surface reactions, nucleation, and growth processes. These early hydration interactions do not appear to have been studied adequately, probably because the pozzolanic reactions are regarded as slow. However, the early hydration reactions determine not only the initial microstructure, but also influence the later strength because the hardened structure is built on the initial framework.

The major effect of the cement on the pozzolanic reaction is the provision of the calcium hydroxide that reacts with the glassy silica to form the calcium silicate hydrate. In addition, we have seen that the alumina in the glass also forms calcium aluminate hydrates, or calcium sulfoaluminate hydrates in the presence of sulfates. As in the case of the hydration of the cement, these reaction rates will depend upon the total concentrations of calcium, alkalies, sulfate, silica, and aluminate in solution from both the cement and the fly ash. The reactivity of the glassy phase depends upon both intrinsic and extrinsic factors, such as the alkali content in solution. High alkali cements should increase the rate of reaction of the glass. Our understanding of these processes has been developed by studies of reactions of pure cement compounds as well as portland cements with fly ashes and pozzolans.

C_3S-Pozzolan-Water Reactions

In studies of the early hydration of C_3S in the presence of fly ash, several workers[3,202] have reported retardation of the main C_3S hydration heat evolution peak. Jawed and Skalny[210] found that two chemically similar Class F fly ashes of different finenesses (314 and 250 m^2/kg Blaine) had very similar retarding effects in water, but quite different effects in 0.5 molar NaOH solutions, as shown in Fig. 8.1. Further experiments showed that C_3S hydration was retarded in the presence of $NaAlO_2$ (50 ppm Al) and that concentrations of Al and Si in solution in NaOH solutions containing fly ash were much higher than in water-fly ash suspensions, appreciably more Al (about 100 ppm) dissolving from the finer fly ash. Furthermore, analysis of the fly ash surfaces indicated that almost all of the surface sulfur was dissolved into NaOH, but not into water. Jawed and Skalny suggested that the retardation was a result of (1) chemisorption of some Ca^{2+} ions on the fly ash particles, which reduced Ca^{2+} concentration and delayed nucleation of calcium hydroxide, and (2) the poisoning effect of soluble silicate and

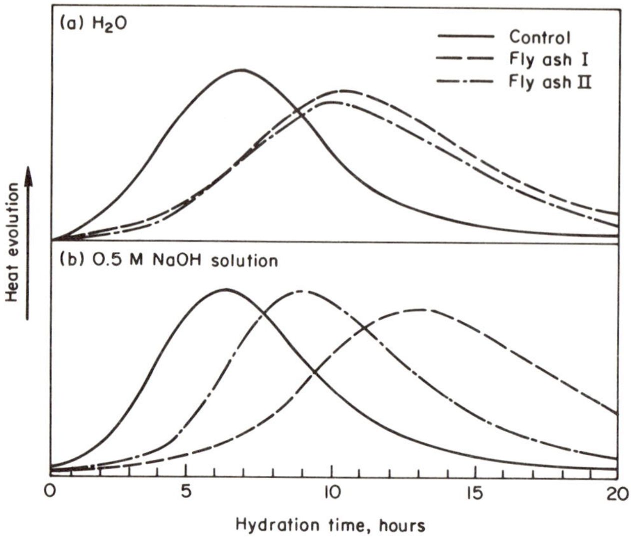

Fig. 8.1 Heat evolution profiles for C_3S-fly ash systems hydrated in (a) H_2O and (b) 0.5 M NaOH solution. (Ref. 210)

aluminate species on nucleation and crystal growth of calcium hydroxide and calcium silicate hydrate.

Differential thermal analysis (DTA) at 8 hours showed that the amount of calcium hydroxide formed during C_3S hydration with the finer fly ash in NaOH solution was less than in water, whereas for the pure C_3S and for mixtures with the coarse fly ash, NaOH accelerated the formation of calcium hydroxide. These DTA results were in apparent agreement with the heat evolution curves. Because of this agreement, it can be tentatively concluded that the amount of calcium hydroxide present depended only on the C_3S reaction rate and that the reaction of lime with the fly ash was not appreciable up to 8 hours. Other workers have concluded that pozzolanic reactions are insignificant for several days or weeks.[3,185,186] This question and retardation of early cement hydration will be reexamined later in the chapter in "Portland Cement-Fly Ash-Water Reactions."

The reaction of C_3S with a Class F fly ash of 373 m^2/kg from 1 to 397 days was studied by Mohan and Taylor.[211] Paste mixes with a C_3S to fly ash weight ratio of 3:1 and a water to solids ratio of 0.5 were hydrated at 25° C. The fly ash accelerated the early hydration of the C_3S with respect to pure C_3S hydration, but after about 28 days their degrees of hydration were nearly equal. The amount of reacted fly ash increased from about 2% at 1 day to 48% at 397 days. The amount of calcium hydroxide increased from about 0.2 g/g C_3S at 1 day to about 0.3 g/g C_3S at 90 days and decreased slightly thereafter, as has been found with portland cement-fly ash pastes (discussed at end of "Portland Cement-Fly Ash-Water Reactions" in this chapter and shown in Fig. 8.8).

Analytical electron microscopy showed that the Ca to Si ratio of C-S-H particles ranged from 1.22 to 1.75 with a mean of 1.43, only slightly below that of 1.51 found for the pure C_3S pastes. It was also found that the degree of polymerization of the silica in the C-S-H in pastes containing fly ash was significantly higher than in pure C_3S pastes after several weeks of hydration.

Takemoto and Uchikawa[185] suggested somewhat more complicated mechanisms for the pozzolanic reaction in the presence of C_3S and C_3A than with calcium hydroxide. For the C_3S-pozzolan system, the process was described as follows and as shown in Fig. 8.2.

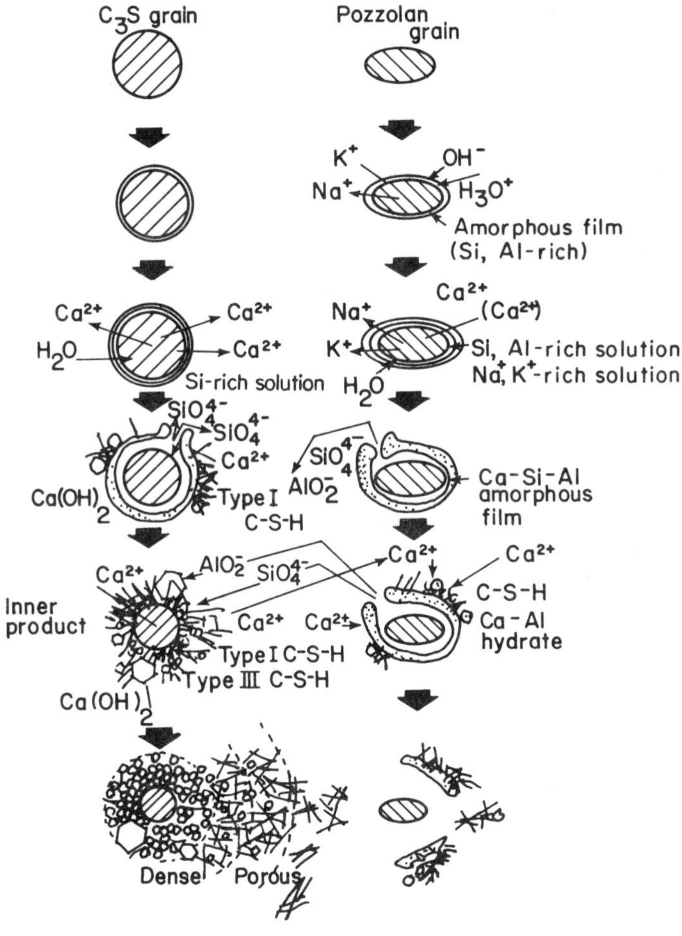

Fig. 8.2 Schematic representation of the mechanism of hydration in the system pozzolan-C_3S. (Refs. 185, 202)

Calcium ions released from the C_3S surfaces saturate the liquid and are adsorbed on the pozzolan particle surfaces. Calcium silicate hydrate formed by the hydration of C_3S precipitates on the surface of the C_3S particles and as porous hydrates of low Ca to Si ratio on the pozzolan grains. Attack of the pozzolan surface in water brings about gradual dissolution of Na^+ and K^+, resulting in Si-rich and Al-rich amorphous layers on the surfaces. Dissolved $Na^+$and K^+ increase the OH^- concentration and accelerate the dissolution of SiO_4^{4-} and AlO_2^-, which combine with Ca^{2+} to increase the thickness of the hydrate layer. This layer swells by osmotic pressure, and a void is formed between the layer and the pozzolan particle. When the pressure in the void ruptures the film, SiO_4^{4-}, AlO_2^-, Na^+, and K^+ ions escape into the Ca^{2+}-rich solution. Additional calcium silicate and aluminate hydrates are formed on the outer hydrates of the C_3S particles and also to a slight extent on the ruptured film. Solids may not form inside the film for some time because the high concentration of alkalies prevents precipitation of hydrates. For pozzolans of low alkali content, Ca^{2+} is able to diffuse into the space inside the ruptured film and to precipitate calcium silicate and calcium aluminate hydrates on the pozzolan particle. In such cases, no space exists between the pozzolan grain and the hydrate. The glasses in fly ashes contain sulfur and relatively large amounts of alkalies, which tend to concentrate in particle surfaces; for these materials void spaces around the particles are often clearly seen, as shown in Fig. 8.3.

Examination of the microstructure by Field Emission SEM with an energy dispersive X-ray analyzer (FESEM-EDX) showed the variation of the Ca to Si molar ratio in several zones. In one typical case, the Ca to Si ratio was nearly constant at about 2.0 in a zone (2 μm) of hydrated C_3S (inner product); outside that zone it decreased for 2 μm to become zero at the surface of a fly ash particle, as shown in Fig. 8.4. The plot of the Ca to Si ratio versus distance shows the excellent resolution of this FESEM-EDX system (Hitachi Seisakusho S-700 with Kevex 7000). In other regions where no pozzolan was present, the Ca to Si ratio at equal distances from C_3S particles remained much higher (about 1.5).

C_3A-Pozzolan-Water Reactions

Takemoto and Uchikawa also examined the hydration of C_3A in the presence of calcium hydroxide, gypsum, and pozzolans. Their concept of the process is depicted in Fig. 8.5.

The presence of the pozzolan accelerates the hydration of C_3A by adsorption of calcium ions from the liquid phase and by providing precipitation sites for ettringite and other hydrates. The general features of the pozzolanic hydration reaction are similar to those of the C_3S system. Ettringite, monosulfoaluminate hydrate, calcium aluminate hydrate, and calcium silicate hydrate are formed on the surface film outside the pozzolan particles or on the surface hydrate layer of the C_3A particles, depending on the concentration of Ca^{2+} and SO_4^{2-} in solution. Ettringite is precipitated only in the early stage of hydration. The SO_4^{2-} concentration in solution is low near the pozzolan particle, andthe hydrate nearest to the grain is mainly the calcium aluminate hydrate. Then the solid solution of C_4AH_{19} with the monosulfoaluminate, and the monosulfoaluminate, form in that order.

Fig. 8.3 SEM micrograph of the fracture surface of hardened C_3S-fly ash paste showing the clearance between the fly ash particle and hydrates. (Ref. 185)

In the case of fly ashes with sulfates concentrated on the surfaces of the particles, these relationships may be different.

Plowman and Cabrera[212] have shown that sulfate was released much more rapidly into solution from a (Class F) fly ash containing 1.25% SO_3 than from solid gypsum (of unspecified fineness) and that the hydration rates of C_3A and C_4AF were controlled much more effectively than with that gypsum.

Portland Cement-Fly Ash-Water Reactions

The studies of the C_3S and C_3A reactions with pozzolans are generally consistent with the more complex reactions between portland cements, fly ashes, and water. According to the workof Dron on the pure compound phase diagram[194] and on lime-pozzolan mortars,[196] gehlenite hydrate should not form in the presence of excess calcium hydroxide. However, the work of Lachowski et al.[179] shows that cement hydration reactions occur under conditions that only approximate local equilibrium. The reaction products of fly ash-cement mixtures are substantially the same as those found in portland cements, i.e., calcium silicate hydrate, calcium hydroxide, calcium aluminate hydrate, ettringite (AFt), monosulfoaluminate hydrate (AFm), and hydrogarnet.[127,201] The amount of calcium hydroxide eventually diminishes with curing time and may be reduced to nearly zero, while the lime-silica ratio of the calcium silicate hydrate tends to be less than in portland cement pastes.

There has been considerable disagreement concerning the time at which the pozzolanic reaction begins. Some workers report no evidence of reaction of the

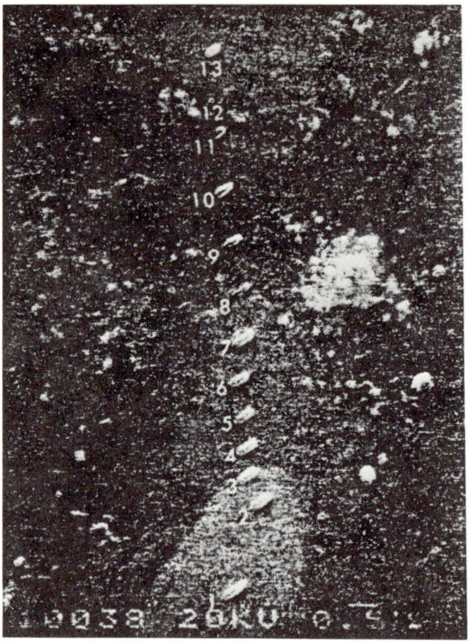

C_3S-fly ash paste, no etching, age 3 days

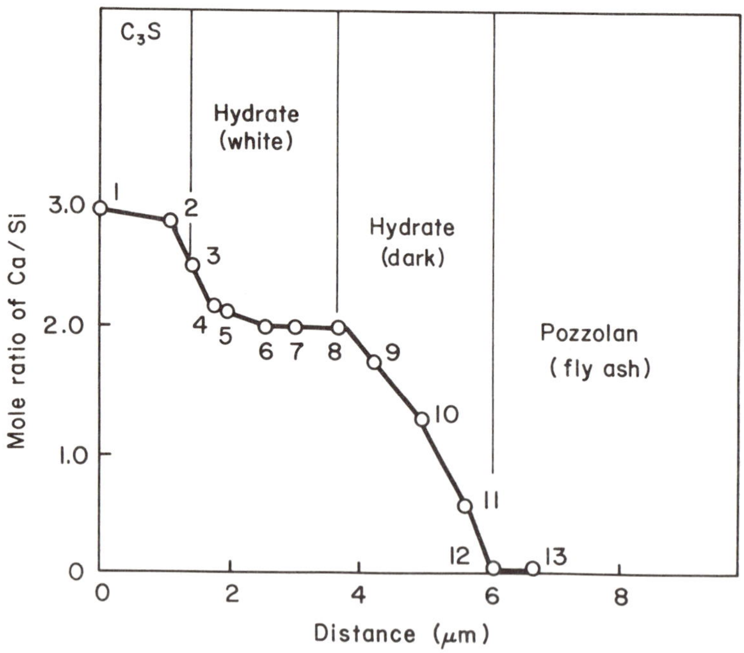

Fig. 8.4 FESEM micrograph of a 3-day-old C_3S-fly ash paste showing the zonal structure of the hydrates and the distribution of calcium and silicon between the fly ash grain and the C_3S grain. (Ref. 185)

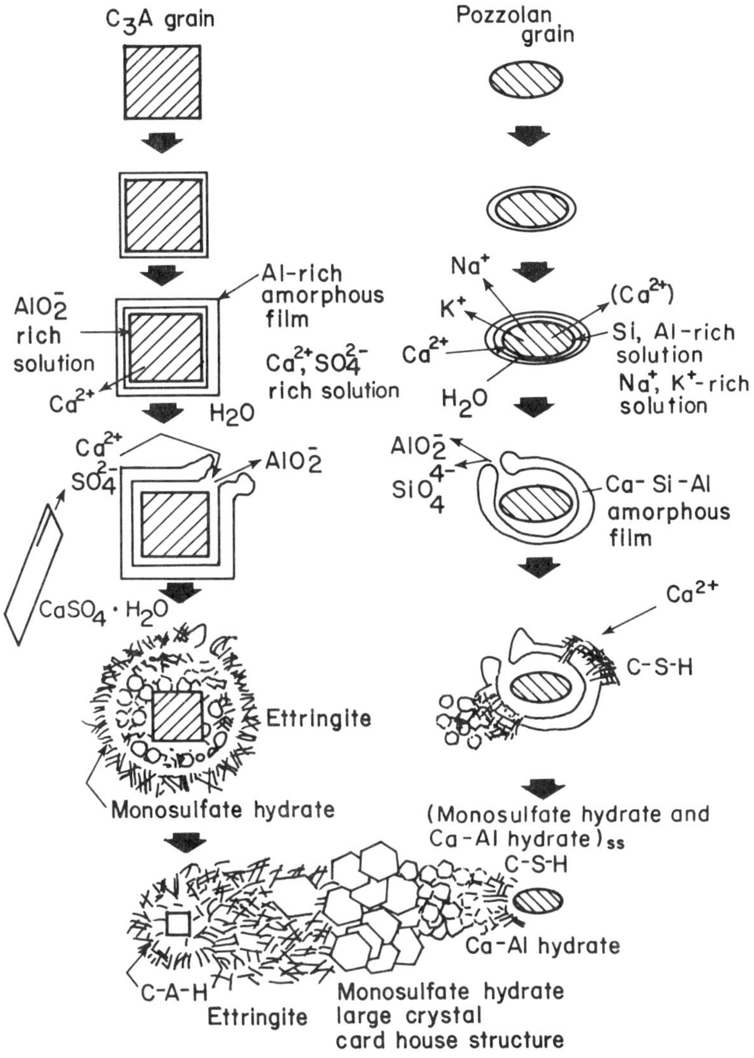

Fig. 8.5 Schematic explanation of the mechanism of hydration in the system pozzolan-C_3A in the presence of $Ca(OH)_2$ and $CaSO_4{\cdot}2H_2O$. (Ref. 185)

fly ash for 28 days or longer, whereas others report very early reactions.[3,186] These differences probably reflect differences in the techniques used to detect reactivity, as well as differences in reactivity. The reaction appears to begin at an early age, but for it to begin to contribute appreciably to strength seems to require some time.

With a highly reactive pozzolan (condensed silica fume) the reaction became significant after several days of curing, and the amount of calcium hydroxide reached a maximum at about 7 days at 20° C, decreasing thereafter.[213] In fly ash cement pastes, more time is usually required to reach the maximum calcium hydroxide content (as in Fig. 8.8).

Takemoto and Uchikawa[185] observed so many similarities between the cement-pozzolan reactions and those of the simpler systems just described that they concluded there was no essential difference between them. With few exceptions, SEM and XRD revealed the same kinds of morphologies, crystalline hydration products, and calcium silicate hydrate. The presence of pozzolans (six different types including fly ash were used) usually accelerated the rate of hydration of the cement, as indicated by the rate of heat evolution at early ages and QXRD of the remaining alite at later ages. With the particular fly ash used, the main silicate hydration heat peak was retarded several hours with both C_3S and portland cement but was accelerated after that. Earlier work has shown that this retardation can be eliminated by washing the fly ash.

Fig. 8.6 shows the rates of heat evolution of portland cement-fly ash-water mixtures during the early reaction period at five different fly ash contents.[3] This is the main (or second) heat peak, which is caused by the acceleration of hydration of alite (C_3S) and is associated with the setting of the cement. In the case of fly ashes washed with water before use, the rate of reaction increased progressively with increasing fly ash content in spite of the decreased cement contents. This is not taken to indicate that the fly ash reacts more rapidly than the cement, but rather that the presence of the fly ash accelerates the cement hydration reaction. This acceleration is a commonly observed effect known as the "finely divided powder effect." Yamazaki[156] explained the effect to be a result of the replacement of cement particles by fly ash (or other fine particles), thereby providing more space in the floc structure for deposition of hydrates by the remaining cement particles. It appears reasonable to suppose that the fly ash particles serve as nucleation and growth sites for C-S-H and other hydrates (see discussion of Fig. 8.7).

Even in the case of a highly reactive pozzolan such as condensed silica fume, Sellevold et al.[213] found that the acceleration of hydration in cement-silica fume pastes at early ages was a result of catalytic action and not a result of pozzolanic reaction.

Also shown in Fig. 8.6 is one curve for the mixture made with the unwashed fly ash, which shows that the heat of reaction was retarded about 2 hours before the maximum rate was reached but proceeded even more rapidly than the washed fly ash after 11 hours of reaction. Similar heat evolution curves were obtained in another study which included both Class F and Class C fly ashes.[214] In addition to the retardation of the heat peaks, the fly ashes retarded the initial set by 2 hours or more and the final set by at least 5 hours. These results are similar to those discussed in "C_3S-Pozzolan-Water Reactions" in this chapter concerning retardation of C_3S hydration by two (unwashed) Class F fly ashes. By contrast, washed fly ash (and condensed silica fume) accelerated the early hydration of portland cements. The acceleration appears to be accounted for in terms of the fine powder effect. Jawed and Skalny[210] attributed the retardation to reduction of the concentration of calcium or increases of the amounts of aluminate and

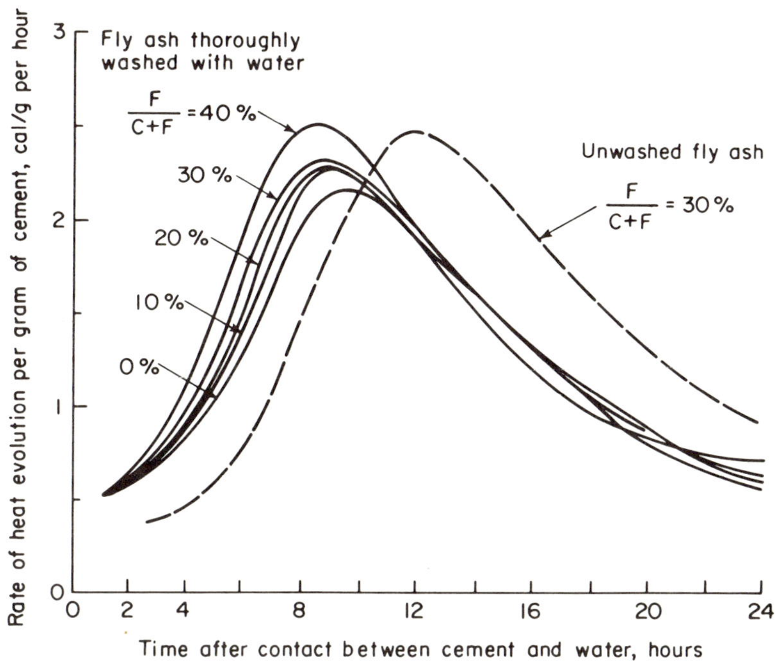

1 cal/(g·h) = 4.186 kJ/(kg·h)

Fig. 8.6 Rates of heat evolution (per gram of cement per hour) in pastes with portions of the cement replaced with fly ash. Water/cement plus fly ash ratio maintained constant at 0.4. (Ref. 3)

Fig. 8.7 Split-screen SEM micrograph of portland cement-fly ash paste hydrated 1 day at room temperature. Magnifications: left photo, 1000X; right photo, 3000X. (Ref. 215)

silicate species in solution. They argued that chemisorption of Ca^{2+} on fly ash particles would *delay* the homogeneous nucleation of calcium hydroxide from solution and thereby *retard* hydration. However, if calcium hydroxide and C-S-H form instead by heterogeneous nucleation on fly ash particle surfaces, *acceleration* would be expected, as has been observed with both washed fly ash and silica fume. Hence, the retardation appears to be a result of some constituent removed from the fly ash by washing. Organics in the fly ash may be such a constituent. Calcium and alkali sulfates are water soluble but are known to be accelerators of alite hydration, despite the fact that they may increase the amounts of aluminate and silicate species in solution. Although the retardation is not yet fully understood, it appears more reasonable to suppose that it is a result of either the adsorption of dissolved organic compounds or the covering of the cement particle surfaces with very fine fly ash particles. These processes were discussed in "Rheological Properties of Fresh Paste" in Chapter 5 to account for the reduced water requirement in mixtures containing fly ash.

Diamond et al.[(215)] have described a morphological aspect of the early reactions which shows the importance of heterogeneous nucleation in fly ash-cement pastes. Prior work had established the common occurrence of so-called "duplex films" which rapidly develop in hydrating cement systems on sand grains and other surfaces in contact with the paste. These films consist of a thin (0.5 μm) layer of calcium hydroxide deposited quickly on the surface plus a thin single layer of parallel, more or less widely-spaced C-S-H gel particles (usually acicular) oriented perpendicularly to the calcium hydroxide surface coating. The combined thickness of the duplex film is only about 1 μm. As hydration proceeds, these duplex films may eventually become bonded by additional hydration products to other particles. It was found that, in 50-50 fly ash-cement pastes hydrated for 1 day at room temperature, such duplex films apparently form on both the fly ash particles and the cement particles, as shown in Fig. 8.7. Subsequent reactions must occur by diffusion through these films. This finding also indicates that the presence of such fuzzy deposits on a particular cement particle in a hydrating paste does not necessarily indicate that the enclosed particle is actively hydrating; inactive cement grains may also collect such deposits as a result of deposition from solution just as do fly ash particles.

In mortars and concretes at later stages of hydration, the oriented calcium hydroxide crystals that formed on the surfaces of aggregate particles show diminishing preferred orientation with distances from the surface of up to 40 μm.[(216)]

The above observations are not entirely consistent with earlier research on fly ash-cement hydration interactions. It is generally agreed that in portland cement pastes calcium hydroxide crystallizes during the first few days in the larger water-filled spaces between the cement grains. Terrier and Moreau refer to this calcium hydroxide as primary portlandite.[(217)] In pastes and mortars containing fly ash or sand, this primary portlandite tends to form instead on the surfaces of the sand or fly ash particles. At later ages (90 days or more) the secondary portlandite begins to form in the interstices between the primary portlandite and the cement grains. The secondary portlandite crystallizes perpendicular to the edges of the grains, forming coatings about 1-μm thick on the fly ash and other particles. Terrier and Moreau suggest that in some specimens thc ash-portlandite contact is generally rather poor and that the appearance of the secondary portlandite in the interstices improves the bond.

These observations of hydrate coatings and duplex films or membranes fit the descriptions of Takemoto and Uchikawa[185] for the previously discussed reaction mechanisms with C_3S and pozzolans. The existence of 0.5 to 1-μm water gaps between such coatings and fly ash particles in fly ash-cement concretes has been documented, and it has been suggested that this is one reason for the delay in strength development.[200] Later strength gains, especially in tensile strength, are associated with the filling of these water gaps with reaction products.

In cement-fly ash mixtures, as opposed to lime-fly ash mixtures, it is more difficult to correlate strength development with degree of reaction because of the need to determine the degree of reaction of each constituent. The degree of reaction of the cement can be determined by QXRD of the remaining alite (C_3S), but the amount of calcium hydroxide must be separately determined, because the C to S ratio of the C-S-H formed by cement hydration may depend on the reactivity of the fly ash.

In the study by Takemoto and Uchikawa,[185] such measurements showed that degree of hydration of the alite in the portland cement-fly ash pastes was higher than in the portland cement pastes from 1 to about 28 days. The amount of calcium hydroxide that reacted with pozzolan was calculated from the difference between the amount actually determined and the amount expected from the degree of hydration, which was determined by QXRD of the alite. These values indicated almost no reaction of pozzolan with the calcium hydroxide for 7 to 28 days for the fly ash and some other pozzolans, despite good strength development. It seems likely that this anomoly is a result of the assumption that the reaction of the pozzolan must reduce the amount of calcium hydroxide because the average Ca to Si ratio of the calcium silicate hydrate is the same, both with and without the pozzolan. This assumption is contrary to the findings of Takemoto and Uchikawa by FESEM-EDX, which showed much lower Ca to Si ratios near the pozzolan particles (see Fig. 8.4 and also "Alkali-Aggregate Reaction Control" in Chapter 11). If the average Ca to Si ratio of the calcium silicate hydrates produced in pastes containing pozzolan is lower than that produced without the pozzolan, the degree of pozzolanic reaction will be higher than calculated by that method. This interpretation is confirmed by examination of results of experiments of the same kind reported by Kondo and Ohsawa,[208] who obtained negative values for the amounts of calcium hydroxide reacted at early ages in fly ash-cement mixtures reacted at 20° C. When calcium hydroxide was used instead of portland cement, positive values were obtained for the amount of calcium hydroxide reacted. It appears that in portland cement-fly ash mixtures the early pozzolanic reaction does not reduce the amount of calcium hydroxide, but instead reduces the Ca to Si ratio of the calcium silicate hydrate.

At later ages it is well known that the amounts of calcium hydroxide produced by hydration in portland cement-fly ash mixtures reach a maximum and then decrease.[7B] Data for cement pastes made with various percentages of a Class F fly ash and moist cured for various periods of up to 14 years are shown in Fig. 8.8.[218] With the 37% fly ash cement, the calcium hydroxide content was reduced to less than 3% in 14 years. The actual rate of consumption of the calcium hydroxide depends on factors other than fly ash content. For very fine pozzolans, such as condensed silica fume, this reduction is much more rapid.[213]

Proportions for Complete Reaction

The above considerations make it apparent that Eqs. 6.1 to 6.6 and 7.4 to 7.6 do not represent exactly the stoichiometry of the hydration and pozzolanic reactions. Nevertheless, it is useful to consider what proportions need to be used for complete reaction of the constituents based on these approximate compositions.[165] This knowledge will lead us to a better understanding of some published research results, standard test methods, and specifications.

Fly Ash-Lime Reactions

The simplest case is that of Eq. 7.4 for the reaction of silica with calcium hydroxide. Let us assume that after complete reaction the average molar lime-silica ratio of the C-S-H is 1.0. This is equivalent to a CaO to SiO_2 weight ratio of 0.93. For the reactions of the alumina in the fly ash with calcium hydroxide, let us calculate the amount of CaO required to fully convert the alumina to gehlenite hydrate as in Eq. 7.6. Gehlenite hydrate also has a lime to silica plus aluminate molar ratio of 1.0, which is equivalent to a CaO to Al_2O_3 weight ratio of 0.55. Additional lime is required to produce AFm and AFt phases. For a fly ash with reactive SiO_2 and Al_2O_3 contents of 50% and 30% respectively (20% other, but negligible CaO content), the amount of CaO required for complete reaction with the fly ash would be

$$0.5 \times 0.93 + 0.3 \times 0.55 = 0.63 \text{ kg CaO/kg fly ash}$$

or

$$1.59 \text{ kg fly ash/kg CaO} = 1.20 \text{ kg fly ash/kg Ca(OH)}_2$$

or

61% fly ash in a fly ash-CaO mixture.

This fly ash content is much less than the value of about 2.0 kg fly ash/kg $Ca(OH)_2$ (2.6 kg fly ash/kg CaO) specified for the test for the Pozzolanic Activity Index with Lime in ASTM C311.[73] The exact value depends upon the specific gravity of the fly ash. Thorne and Watt[219,220] also used much higher amounts, 2.5 kg fly ash/kg CaO, in their tests of pozzolanic activity of fly ashes. In their tests these proportions were selected as those giving maximum strength with very low water content. In tests by Minnick[122] and Joshi and Rosauer,[204] the ratio of ASTM C311 was used. In the ASTM test, the amount of fly ash used is about 65% more than the amount estimated for complete reaction of the lime. Hence, the pozzolanic activity of the fly ash in these tests is being evaluated in a very lime-deficient, silica-rich situation in which the composition and distribution of the reaction products may be very different from those in actual use with portland cement in concrete.

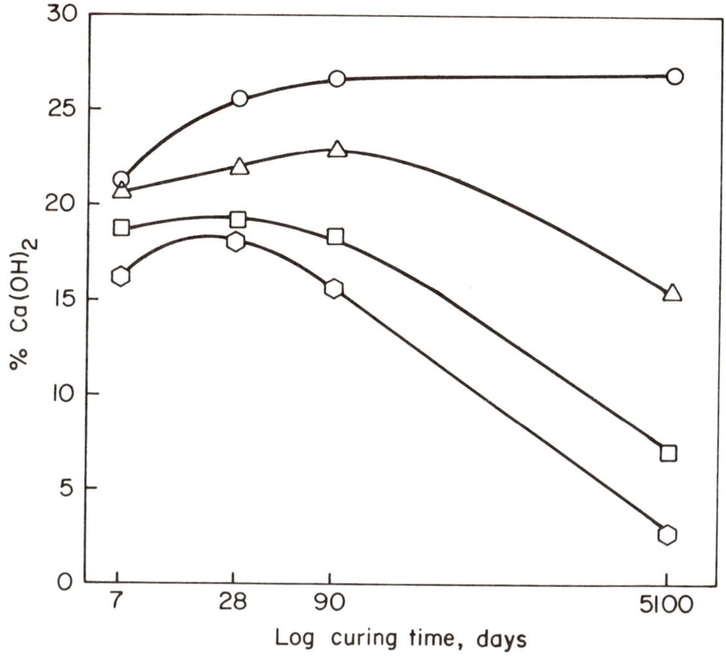

Fig. 8.8 Changes of calcium hydroxide contents of portland and pozzolanic cement pastes during 14 years moist-curing. Calcium hydroxide content by thermogravimetric analysis is expressed as percent of portland cement ignited weight. (Ref. 218)

○ Portland cement paste (LTS 14, high alkali Type I, PCA Research Department Bulletin 26)
△ 16.5% Class F fly ash (Chicago)
□ 28% Class F fly ash (Chicago)
⬡ 37% Class F fly ash (Chicago)

Fly Ash-Cement Reactions

The amount of fly ash required to react completely with the calcium hydroxide produced by hydration of portland cement in cement-pozzolan mixtures may be similarly estimated.[165] At complete hydration, ordinary portland cements yield about 0.20 kg $Ca(OH)_2$/kg hydrates or 0.24 kg $Ca(OH)_2$/kg cement. If we assume the same fly ash composition as in the reaction with lime (50% SiO_2 and 30% Al_2O_3), the amount of fly ash required for complete reaction is calculated to be 0.29 kg fly ash/kg cement, i.e., 22% fly ash.

This estimate is based on cement hydration according to Eqs. 6.1 to 6.6 and reaction of the calcium hydroxide with fly ash to produce C_2ASH_8 and C-S-H with a C to S ratio of 1.0. However, assuming instead that the cement also hydrates to the low-lime products, complete reaction is obtained with 0.50 kg fly ash/kg cement, or 33% fly ash.

These two proportions (22 and 33% fly ash) bracket the proportions specified in ASTM C311 for the Pozzolanic Activity Index with Portland Cement, which specifies that 35% of the volume of the portland cement be replaced with an equal volume of pozzolan. For a fly ash specific gravity of 2.5, this is equivalent to 0.43 kg fly ash/kg cement, or very nearly 30% fly ash and 70% cement by weight. Whereas the proportions specified for the test with lime are estimated to be very lime deficient, the proportions for the pozzolanic activity index test with portland cement appear to be accurate stoichiometrical approximations.

The balance between the total amount of CaO and the amounts of SiO_2, Al_2O_3, and Fe_2O_3 required for complete reaction and maximum utilization of the cementitious properties of the mixture was recognized in the former Italian specifications for portland-pozzolan cements.[2] (Later Italian specifications utilize the pozzolanicity test; see "Pozzolanicity Test" in Chapter 10.) The lowest quantity of pozzolan to be used with a cement was specified by the following total composition ratio for the cement-pozzolan mixture:

$$\frac{\%SiO_2 + \%Al_2O_3 + \%Fe_2O_3}{\%CaO} = 1 \qquad (8.1)$$

If we take for typical bituminous coal fly ash the composition given by Sersale[127] in Table 3.1 and a cement composition with 64% CaO, 21% SiO_2, 5% Al_2O_3, and 2% Fe_2O_3, this ratio equals 1.0 with a 30% fly ash-70% portland cement blend, or 0.43 kg fly ash/kg cement. This figure agrees exactly with the proportions specified for the test for the Pozzolanic Activity Index with Portland Cement in ASTM C311 for fly ash of specific gravity of 2.5.

The ratio used by Takemoto and Uchikawa[185] was even higher, 1.5 kg fly ash/kg cement (60% fly ash and 40% cement). This is about four times the amount we have just estimated to be required for complete reaction.

Variations of Compositions of the Hydrates

There are several good reasons for using somewhat more fly ash in tests or in practice than estimated from the simple reaction equations used here. These estimates of the amount of fly ash that can be completely reacted with lime or portland cement are based on assumed reactive SiO_2 and Al_2O_3 contents of the fly ashes at about the maximum values in Table 3.1. The data presented in Chapter 3 show that the glass contents of representative fly ashes ranged from about 60 to 90%. The material in the most common crystalline phases may not be reactive, and material in very large particles or agglomerates may not be available for reaction. Hence, these estimates of the amounts of fly ash required may be regarded as lower limits; actual values depend on glass content and particle size and may reach values twice as high as these estimates, even with low-lime content fly ashes of the assumed composition. If the CaO content of the fly ash is small but not negligible, as assumed in the above estimates, proportionately more fly ash will have to be used to consume any given quantity of lime completely.

If the reactive CaO content is so high that the silica and alumina contents are substantially reduced, the amount of fly ash required to react completely with the external lime increases more than the proportionate increase in lime content of the fly ash. For such fly ashes, very large ratios of fly ash to portland cement are required for the assumed stoichiometry. In this connection it is important to note that for a lignite fly ash of composition (cf. Table 3.1) 25% SiO_2, 12% Al_2O_3, 8%

Fe_2O_3, and 45% CaO, the former Italian specification of Eq. 8.1 could not be met with any composition of less than 100% fly ash.

These matters are more clearly illustrated in Fig. 8.9, which was constructed from calculations of the amounts of CaO required to react completely with fly ash of the typical bituminous coal fly ash composition used above. The quantity

$$\frac{C - S}{A + F} = \frac{C - C'}{A + F} \tag{8.2}$$

is the average molar C to (A + F) ratio of the sum of the C_2ASH_8, AFm, AFt, and CH hydrate phases which can be formed with C-S-H with a C to S ratio of 1.0. If the fly ash is 100% reactive, the curve on the left indicates the variation of the aluminoferrite hydrate C to (A + F) ratio with the percent fly ash. This steeply descending curve shows that when the blend exceeds 61% fly ash, the (C − C′) to (A + F) ratio is less than 1.0, i.e., there is not enough CaO to produce both C_2ASH_8 and C-S-H with C to S = 1.0. This composition is below the region of C-S-H stability indicated in Fig. 7.2 or 7.4. At lower fly ash contents there is enough CaO to produce the higher lime AFm and AFt phases. It is assumed that when AFt phases are present, calcium hydroxide (CH) will also be present.[198] The upper bound of the C-S-H + AFm + AFt + CH region is uncertain, because the C to S ratio of the C-S-H may exceed 1.0 and CH may remain. The amount of sulfate is important in this compositional balance.

The calculation was repeated assuming that only 80% or 60% of the fly ash was in reactive glassy phases of sufficiently small particle size to react. These curves show how the amounts of fly ash must increase for complete reaction with the CaO. Alternatively, the curves show how the expected products (or reaction chemistry) change with the same fly ash contents. The point of complete reaction with CaO at 61% fly ash is then seen to extend upward with decreasing fly ash reactivity, traversing the range for C-S-H + AFm + C_2ASH_8 phases and then the range for C-S-H + AFm + AFt + CH phases. By contrast, the proportions specified by ASTM C311 (72% fly ash) show that only with a fly ash that is less than 60% reactive would complete reaction result in a C-S-H with C to S = 1.0; a more reactive fly ash would result in a lower C to S ratio hydrate, or hydrated silica.

If the reaction reduces the calcium hydroxide content to below the range of stability of the monocalcium silicate hydrate, decomposition of the C-S-H should begin to occur. Since the principal cementing component is the calcium silicate hydrate, its decomposition may adversely affect the mechanical properties of test specimens.

If Eq. 8.1 is used to determine the fly ash to CaO ratio for a pozzolanic mixture indicated by the former Italian specification, we obtain a 55% fly ash mixture for the 50% SiO_2-30% Al_2O_3 fly ash considered. This point falls near the upper boundary of the C-S-H + AFm + C_2ASH_8 range. This result is consistent with the objective of having sufficient pozzolan present to have no remaining calcium hydroxide.

Calculations for complete reaction of fly ash-portland cement blends cannot be indicated on the same diagram. However, an equivalent calculation was made

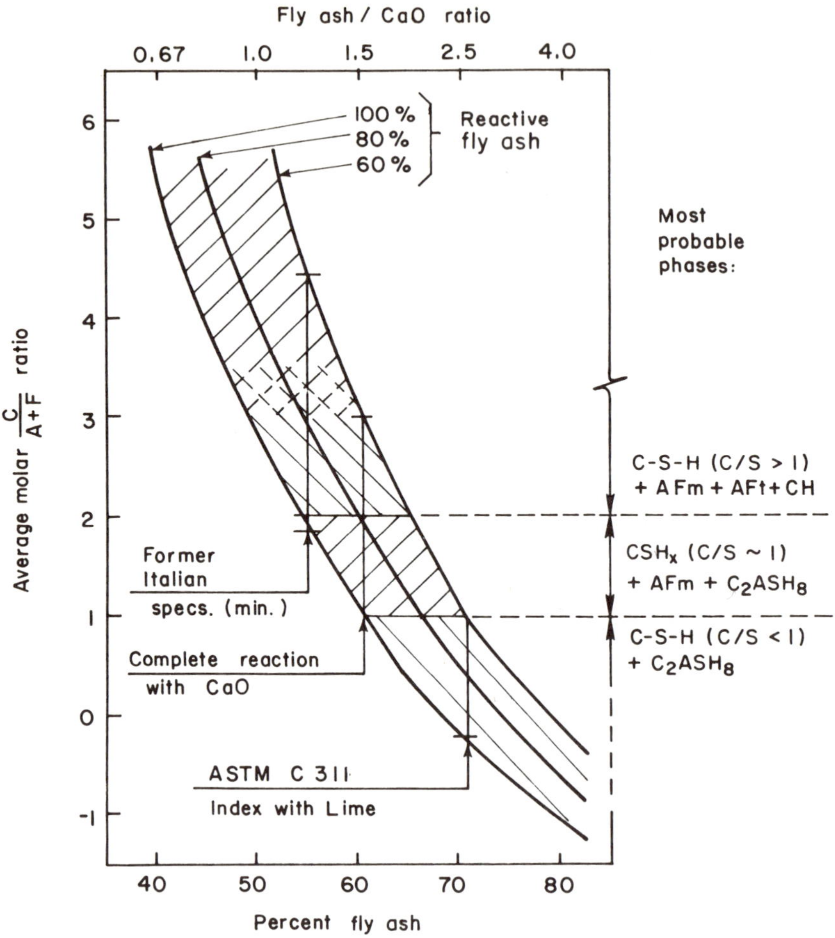

Fig. 8.9 Calculated average molar C/(A + F) ratio of the sum of the hydrates (C_2ASH_8, AFm, AFt, and CH) which can be formed with C-S-H of C/S = 1.0 during complete reaction of CaO in CaO-fly ash-water mixtures. (Fly ash assumed to be 50% SiO_2, 30% Al_2O_3.) Most probable phases are indicated for each range of composition. (Ref. 165)

for the 30% fly ash-70% portland cement mixture indicated in both the Pozzolanic Activity Index with Portland Cement in ASTM C311 and in the former Italian specification for portland-pozzolan cements. For the typical compositions used, this mixture corresponds to a (C - S) to (A + F) ratio of 2.2, which is near the lower end of the C-S-H + AFm + AFt + CH range. This is in an entirely different composition range from that of ASTM C311 for the Pozzolanic Activity Index with Lime.

Consequences of Fixed Proportions

It is apparent from the foregoing that if a fly ash has a high silica-alumina reactive glass content, a low lime content, and a high fineness, the pozzolanic reaction will be limited by lack of lime in the ASTM C311 test with lime but will probably have a slight excess of lime in the test with portland cement for typical compositions. Use of such fixed proportions in standard tests has two important implications:[(165)]

1. Much more fly ash will remain unreacted in the ASTM C311 test with lime and similar tests than in comparable tests with portland cements or in concrete using ordinary fly ash to cement ratios. The hydration products will also be different.
2. Fly ashes compared in the ASTM C311 test with lime and similar tests will be compared on the basis of the reactivity of a relatively small fraction (the most rapidly reacting fraction) of the ash. This limitation must also have the effect of reducing the differences in performance between fly ashes containing some minimum amount of reactive glass and others containing much more. The more reactive fly ash could conceivably yield poorer test results.

Examples of low strengths in pozzolanic activity index tests with lime that may be attributable to inadequate lime contents were reported in a study of 33 fly ashes by Brink and Halstead.[(143)] Their test method with lime was similiar to that in C311. They also prepared cement-fly ash mortars similar to those of C311 but cured at 73° F (23° C) to simulate the use of fly ash in concrete in practice. Comparison of the two sets of strength tests showed that six of the fly ashes did not follow the general trend. Five of these ashes showed unexpectedly low strength when tested with lime. These five ashes were among the nine best (highest strengths with cements) ashes of the 34 ashes. Although the differences in water requirements may have accounted for some discrepancies, it appears likely that lime deficiency in the tests with lime was responsible for most of the low strengths.

The results shown in Fig. 1.3 for decreases in strength with increasing pozzolan contents (after reaching a maximum strength at an optimum pozzolan content) may be examples of the same effect in concretes.

Whatever fixed proportions are used, the conditions of test and use will be optimized for only a narrow range of fly ash compositions, glass contents, and fineness. It appears that tests would be more meaningful and superior performance would be achieved by comparisons at proportions optimized to result in the desired composition of the hydrated phases.

Porosity of the Hardened Cements

Other factors influence strength and rate of strength development in addition to pozzolanic reactivity. Among the most important is the porosity, which is determined by the initial water content of the mixture and by the extent of the hydration and pozzolanic reactions. The pozzolanic reactions are important because they contribute to the space-filling process performed by hydration of the cement.

Takemoto and Uchikawa have compared the porosity and pore structures of portland cement pastes of 0.3 water-solids ratio with portland-pozzolan cement pastes made with 40% portland cement.[185] Pastes were cured for periods of up to 180 days and determinations made of total porosity, capillary pore space, gel pore space, and compressive strength.

In these and other similar comparisons, the initial porosities are affected by the difference between the specific volume of the cement (about 0.317 cm^3/g) and that of the pozzolan. In this case, the Takehara fly ash had a specific volume of 0.413 cm^3/g. If we assume that the air content is small, the volume of paste, V_p, in these mixtures can be calculated from

$$V_p = w_0 v_w + c v_c + f v_f \tag{8.3}$$

in which w_0, c, and f are the masses of the water, cement, and fly ash, and v_w, v_c, and v_f are their specific volumes. For the portland cement paste, the yield was 0.62 cm^3/g of cement; for the portland-pozzolan pastes, the yield was 0.67 cm^3/g of solids.

The initial porosity p_0 of these mixtures can be calculated from

$$p_0 = \frac{w_0 v_w}{w_0 v_w + c v_c + f v_f} \tag{8.4}$$

From this equation, the initial porosity of the cement paste was found to be about 0.48, whereas that of the cement-fly ash paste was only about 0.44. Although the cement-fly ash paste had a larger volume, it had a lower porosity. These calculations, however, neglect some of the porosity of the fly ash itself.

In both cases, the hydration and pozzolanic reactions caused the total and capillary porosities to decrease steadily while the gel pore volume steadily increased, as shown by the values given in Table 8.1. For the portland cement paste, the total porosity decreased to about 0.40 in 1 day and to about 0.20 in 180 days. The capillary space decreased from about 0.15 at 3 days to about 0.02 at 180 days.

For the pastes containing fly ash, the total porosity decreased to about 0.35 in 1 day and to about 0.24 in 180 days. The capillary space decreased from about 0.20 at 3 days to about 0.04 at 180 days. The volume of gel pores increased more rapidly at early ages in the portland cement pastes, but after 91 days the volume of gel pores was greater in the fly ash paste. These measurements show that, although the portland cement was a more effective space filler, the fly ash was also quite effective in reducing total and capillary porosity in these mixtures.

One of the reasons fly ash is not as effective a space filler as portland cement is that some of the fly ash particles are themselves porous. The spongy glass particles, cenospheres, and sintered agglomerates each contain large pores. In addition, the coke particles contain fine pores. Hence, although our initial calculation indicated that the initial porosity of the paste with fly ash was lower than that of the portland cement paste, some additional porosity was present in the fly ash. During the course of the reactions, at least some of this pore space is opened up and provides additional space to be filled with reaction products.

Table 8.1 Porosities of Portland Cement and 40% Portland Cement with Fly Ash Pastes at Various Curing Times,[185] water/solids = 0.3

Curing time, days	Portland cement paste porosities			Fly ash + 40% portland cement paste porosities		
	Gel	Capillary	Total	Gel	Capillary*	Total*
0	0	0.48	0.48	0	0.44	0.44
1			0.40			0.35
3	0.134	0.15	0.28	0.113	0.20	0.32
91	0.150			0.201		
180	0.180	0.02	0.20	0.20	0.04	0.24

*Some of the porosity of the fly ash particles was not detected by the measurements.

The pore structures of these pastes were examined in more detail by two different kinds of pore size distribution measurements. The capillary space referred to above was determined by mercury porosimetry to be that corresponding to pores larger than 0.015 μm (15 nm) in diameter. At 3 and 7 days of curing, the portland cement pastes and the pastes with fly ash had similar pore size distribution curves with two broad maxima at about 0.02 and and 0.2 μm. Further curing greatly reduced the peak at 0.2 μm in the portland cement paste but had little effect in further reducing these pores in the paste with fly ash. However, in both pastes additional curing substantially reduced and shifted the peak at 0.02 μm to smaller sizes.

Feldman[221,222] has reported somewhat similar results for mixtures of compositions (65% cement and 35% fly ash) closer to those used in practice. However, he found that pore size distributions obtained by high-pressure mercury intrusion porosimetry for the fly ash-blended cement pastes were incorrect because of damage to the pore structure during intrusion. It was also found that the total porosity of the hydrated blended cements could not be determined by saturation with methanol or helium pycnometry. This result was distinctly different from results obtained with portland cement pastes, which yielded equal porosities by all three methods. However, reasonable values for the total porosity of the blended cement pastes were obtained from the porosimetry data at 408 MPa (60,000 psi). It was concluded that in the blended cement pastes some of the larger pores were blocked by hydrates until disrupted at high mercury intrusion pressures. In portland cement pastes the pores are not blocked, and so methanol, helium, and mercury all penetrate the same pore volumes.

The gel pore size distributions were determined by Takemoto and Uchikawa from nitrogen adsorption isotherms. These measurements showed that the gel pore peak at about 2 nanometers (0.002 μm) increased much more for the fly ash paste than for the cement paste between 3 and 91 days. The gel pore volume of the cement paste increased from 0.134 to 0.150 in from 3 to 91 days; for the paste with fly ash the increase was 0.113 to 0.201, which indicates that the porosity of the gel in the pastes with fly ash is considerably higher than that of the cement gel. The higher gel porosity is consistent with the fact that the portland cement gel also contains crystalline calcium hydroxide; in the pastes with fly ash the nonporous calcium hydroxide reacts to produce more of the porous calcium silicate hydrate gel. However, the higher gel porosity may simply be a result of the fact that the pozzolanic reaction ceased before the minimum gel porosity was reached. With the proportions used (40% cement), the overall ratio of (SiO_2 + Al_2O_3) to CaO was almost 2.0, indicating that the supply of lime from the cement hydration was not nearly adequate to react with all of the pozzolan.

It is apparent that although the space-filling processes and pore structures produced in cement-fly ash pastes are very similar to those in portland cement pastes, there are also some significant differences. Portland cement particles are denser, essentially nonporous, concentrated solids which are able to fill more capillary space with denser cement gel than do pastes containing fly ash, even when the pastes with fly ash are initially less porous. This is at least partly because the porosity of the fly ash provides more space to be filled by hydration products. These factors tend to reduce the strength development in fly ash-containing pastes. However, the strength data showed that both the portland cement and fly ash mixtures reached very high strengths (100 and 90 MPa at 91 days, respectively). The slightly lower strengths for the fly ash mixtures are consistent with their higher porosities, but the reduction is small considering the large amount of cement replaced. If a somewhat higher proportion of cement had been used, or if the water content of the fly ash mixture had been reduced to yield a smaller volume of paste (as for the portland cement pastes), the 91-day strengths of the fly ash mixture would probably have exceeded those of the portland cement pastes.

Strength Development

A pictorial representation of the processes producing strength development is shown in Fig. 8.10. Based on a study of the hydration of oil well cement (API Class H) mixtures containing high lime (Class C) fly ashes, the structure formation processes were monitored by SEM and strength tests.[223] The hydration was monitored by XRD. During the early curing period, the bonding of the fly ash particles to cement matrix by radiating bundles of fibrous C-S-H is very weak, but channels in the C-S-H close with time. As the thickness of the fly ash hollow shell is reduced by reaction, the C-S-H network becomes thicker, denser, and stronger so that at later ages the structure fails by fracture through the fly ash shells, rather than by pullout.

There are various ways to develop quantitative relationships between the strength of cementitious systems and their porosity, or the extent of filling of the pore space with hydration products. Powers worked out these relationships in terms of the gel-space ratio, X, which is the ratio of the volume of the hydrated cement gel of minimum porosity to the space available for gel formation.[181] The

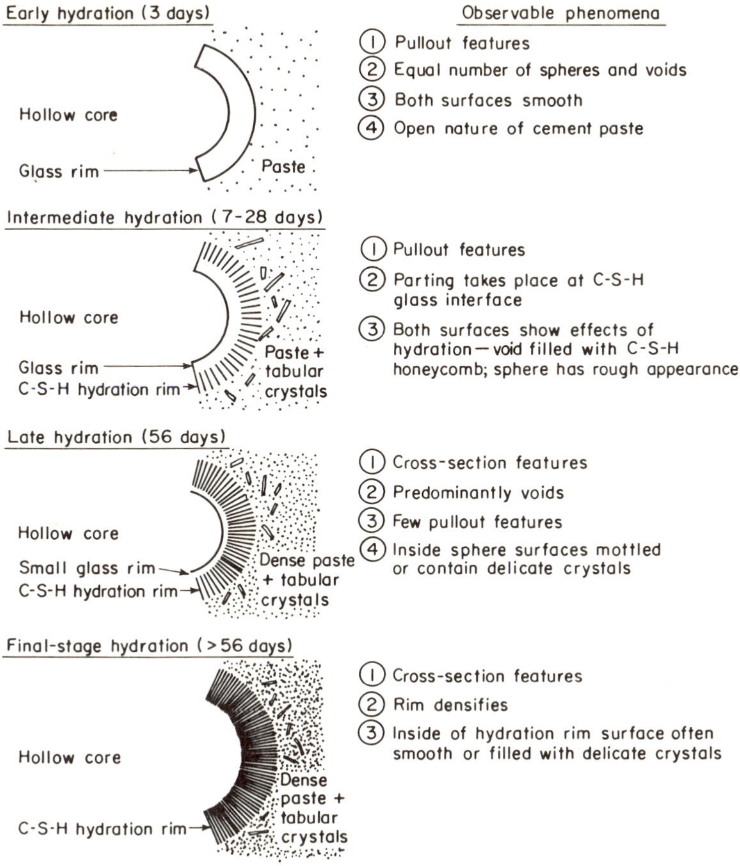

Fig. 8.10 A model of high-lime fly ash hydration. (Ref. 223)

volume of gel at any stage of hydration is proportional to the degree of hydration and is about 2.2 times the volume of the cement before hydration. The space available depends mainly on the water-cement ratio. Then the compressive strength, S, of hardened cements approximately fits the power law

$$S = S_0 X^n \tag{8.5}$$

in which S_0 is the intrinsic strength of the densest ($X = 1$) gel produced by a given cement under normal hydration conditions and n has a value of about 3.

Takemoto and Uchikawa[185] used a different function of the porosity, p, to fit the data obtained with cement pastes and pozzolanic cement pastes:

$$S = S_0 e^{-bp} \tag{8.6}$$

It should be noted that Eq. 8.6 cannot be correct at high porosities because the strength must actually approach zero as p approaches unity. Moreover, Feldman's data show that the equation does not apply over the range of porosities of interest.[221] It may be appropriate, however, for use with pastes compacted under pressure to have finite initial strength.

It is also of interest to compare the strength development of fly ash-lime mortars tested by Watt and Thorne [219,220] with the fly ash reaction rates shown in Figs. 9.1 and 9.2. Fig. 9.1 shows the degree of reaction (percent) calculated from a simple model of spherical particles with equal surface reaction rates. Fig. 8.11 shows the strength data for these same fly ashes with median diameters ranging from 7 to 46 μm. There is very good general agreement between these strength data and the calculated reaction rates, although Watt and Thorne noted that not all of their results showed as much agreement as did these five samples.

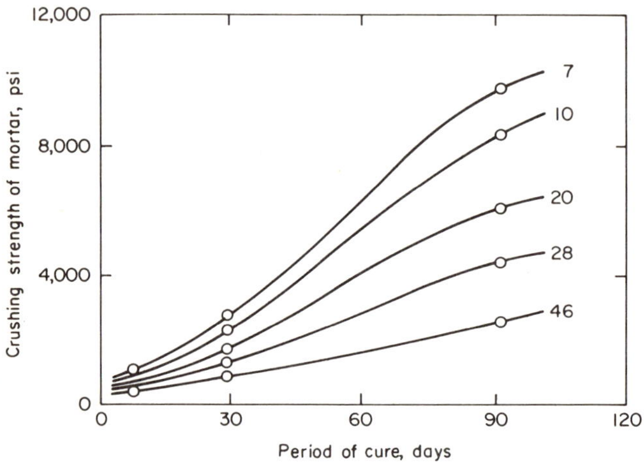

1000 psi = 6.894 MPa

Fig. 8.11 Mortar strength development for fly ashes at various median particle diameters (as indicated, in microns). (Ref. 220)

Actually, there is no reason to expect a simple proportional relationship between strength and degree of reaction. Strength depends on initial porosity and degree of reaction, or degree of space filling, and intrinsic strength of the gel. Powers' use of the gel-space ratio, as in Eq. 8.5, as a power law shows that the later stages of the reaction contribute much more to strength development than does an equal amount of reaction early in the process. In Watt and Thorne's results, the later strengths may not have been as high as expected on this basis because of the porosity of the larger fly ash particles and because their mortars had been compacted under high pressure to yield substantial initial strengths.

It should also be noted that Watt and Thorne used a fly ash to CaO ratio of 2.5. By our calculations in "Proportions for Complete Reaction" in this chapter, these mortars might be thought to be too lime deficient to permit complete reaction of the fly ash and normal hydration products. However, because of their very low porosities, all of the available space would be filled before the fly ash could be completely reacted even if lime were present. Hence, the reaction products were probably in the range of those expected in normal proportions in concrete, although unreacted fly ash, and perhaps also lime, remained.

The effects of changing the particle size distributions of fly ashes on strength and flow of cement-fly ash pastes were examined by Montgomery.[151] The properties of pastes made with fly ashes which had been classified by particle size and reconstituted to a single particle size distribution were compared with the properties of pastes made with the as-received fly ashes. The differences in paste strengths with the reconstituted fly ashes were smaller than those for the as-received ashes, as were the average coefficients of variation.

In this chapter we have examined in some detail interactions between cement hydration and pozzolanic reactions. We have seen how the proportions used affect the nature of the products, the initial porosity, and how the porosity and pore structure change during the reactions. Although some fundamental considerations concerning rates of early reactions and development of microstructure and strength have been outlined, we have not yet synthesized information concerning characteristics of fly ashes that influence pozzolanic reactivity. In the next chapter we will review how measurable characteristics of fly ashes directly affect pozzolanic reactions and additional work on how strength tests are used to measure reactivity.

Chapter 9

Characteristics Influencing Pozzolanic Reactivity

Definitions

The term "pozzolanic reactivity" is not well defined. The definition used in "Definitions" in Chapter 7 as the degree of reaction is evidently inadequate unless a reaction time and test condition are specified; if these are specified, the term implies an average rate of reaction. If the specified reaction time is very long, the pozzolanic reactivity could be taken to mean the ultimate degree of reaction. In fact, the term "degree of reaction" implies the ability to determine the 100% reacted state, or the ultimate degree of reaction, which may present practical difficulties. In addition, it may be difficult to determine, or even define, the "degree of reaction" because there are several different reactions taking place and these may occur at different rates.

The definition of the reactivity of a fly ash[(186)] as the rate at which strength-giving phases are produced by the reactions between the fly ash and a portland cement in the presence of water comes closer to generally accepted usage but also suffers from some of the same inadequacies. The rate of reaction depends on test conditions in addition to fly ash characteristics. Also, changes in the test conditions may alter the reactions so that different reaction products or different forms occur.

For these reasons many tests of fly ashes are "pozzolanic activity" tests of strength development, as in ASTM C311, rather than rates of chemical reaction. Some of this strength data is included in this section, but most will be considered later in "Pozzolanic Activity and Strength" in this chapter. There are factors other than chemical reaction rates which also affect strength development, as was discussed in "Proportions for Complete Reaction" and "Strength Development" in Chapter 8.

Clifton et al.[(186)] distinguished between the intrinsic and extrinsic factors influencing reactivity of fly ashes with lime and cement. The intrinsic factors include chemical and mineralogical compositions, physical properties, and the variations in properties from particle to particle. Extrinsic factors include such factors as the mix proportions, lime or cement compositions, and curing conditions.

Most Important Characteristics

Most reviewers have agreed that the characteristics of fly ashes that most strongly influence the pozzolanic reactivity are fineness, glass content, and silica or silica-plus-alumina content. Other factors may also influence reactivity and especially strength development resulting from the reactions. Most published data involve the effects of several parameters changing simultaneously so that it is seldom possible to obtain a high correlation coefficient with any single variable except fineness.

Fineness

A theoretical relationship between the rate or degree of reaction of fly ash with lime and the fineness of the fly ash was developed to compare with the amount of reacted material determined chemically.[220] A simple model of the reaction process was used to calculate relative rates of reaction from the particle size distribution curves. The following assumptions were made:

1. All the fly ash particles are spherical, each with an initial radius r_0.
2. d is the thickness of the reacted surface layer after time t, all parts of the surfaces being attacked at the same rate so that $d = kt$. Then the reacted volume of any fly ash particle was calculated from

$$V = -\int_{r_0}^{r_0 - d} 4r^2 dr \tag{9.1}$$

The rates of reaction with lime of the 14 well-characterized fly ashes used by Watt and Thorne[123] were experimentally determined using the Feret-Florentin Method.[224] After curing for various times, lime-fly ash compacted mortar specimens (see "Thorne and Watt's Mortar Tests" in this chapter for details) were extracted with HCl; the extract was then analyzed for (SiO_2 + Al_2O_3 + Fe_2O_3). The unreacted fly ash was similarly extracted and analyzed. The difference between these two analyses is taken to be the amount of pozzolanically active material that has been made soluble by reaction with lime. In a modification of the Feret-Florentin Method, called the Lime Solution Method, 0.5 g fly ash was reacted with 0.5 g CaO in 50 mL water and extracted as in the mortar method.

The results of relative rates of reaction calculations using Eq. 9.1 are shown in Fig. 9.1 for five fly ashes with median diameters ranging from 7 to 46 μm. These results may be compared with experimental measurements of the degree of reaction as determined by the acid-solubles increment using the Lime Solution Method shown in Fig. 9.2. The general agreement between the calculated and experimental curves, both in form and in relative position, confirms the assumptions on which the calculations are based. The experimental curves, however, suggest that the fly ashes with larger median particle sizes are more reactive than would be expected from the model, perhaps because the larger particles are predominantly spongy. Microscopic observation showed that they perforate during reaction with lime.

Statistical Correlations

Takemoto and Uchikawa[185] used linear regression to analyze the effect of four characteristics on the amount of calcium hydroxide that was reacted in tests of six different pozzolans, including one fly ash. The four characteristics chosen as independent variables were wt. % glass, wt. % HCl soluble, the silica to alumina ratio in the acid soluble part, and the BET specific surface area. The fly ash was

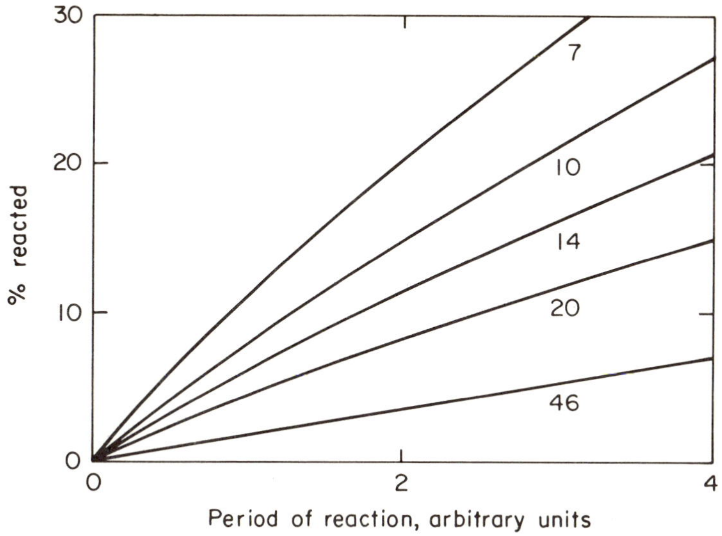

Fig. 9.1 Calculated reaction rates of typical fly ash of various median particle diameters (as indicated, in μm). (Ref. 220)

48% soluble in hot HCL. The weighting factors in the linear regression equation were found to vary erratically with reaction time, with temperature, and with reactant (calcium hydroxide or cement). The fact that high correlation coefficients were obtained at each temperature cannot be considered significant, because only six materials were used with five adjustable parameters to fit the data. Furthermore, it was found that the correlation between the amount of calcium hydroxide reacted and strength development varied substantially with curing temperature and mix proportions.

Thermal History and Chemical Factors

It has sometimes been suggested that residual strain in the glass may also contribute to reactivity of fly ashes. It has been shown, however, that such strains are negligible for spheres less than 1 mm in diameter and become significant only for spheres greater than 10 mm in diameter.[225] Residual strains caused by partial crystallization appear to be important[137] but difficult to quantify.

Ings[226] used very carefully prepared, equal-sized (2 μm) fractions that were nearly 100% glass which had been separated from five different bituminous coal fly ashes to determine what factors, other than fineness and glass content, influenced reactivity. Primary emphasis was placed on factors related to the thermal history of the fly ash, such as OH content of the glass, iron oxidation state, and fictive temperature (as determined from density changes upon annealing). The fictive temperature is the temperature at which the glass structure frozen in during cooling would have been the equilibrium structure. The early reaction rates of all of the glasses were very slow in limewater, ranging from about 16 to 35% reacted in the first 50 days, despite the small particle size. The

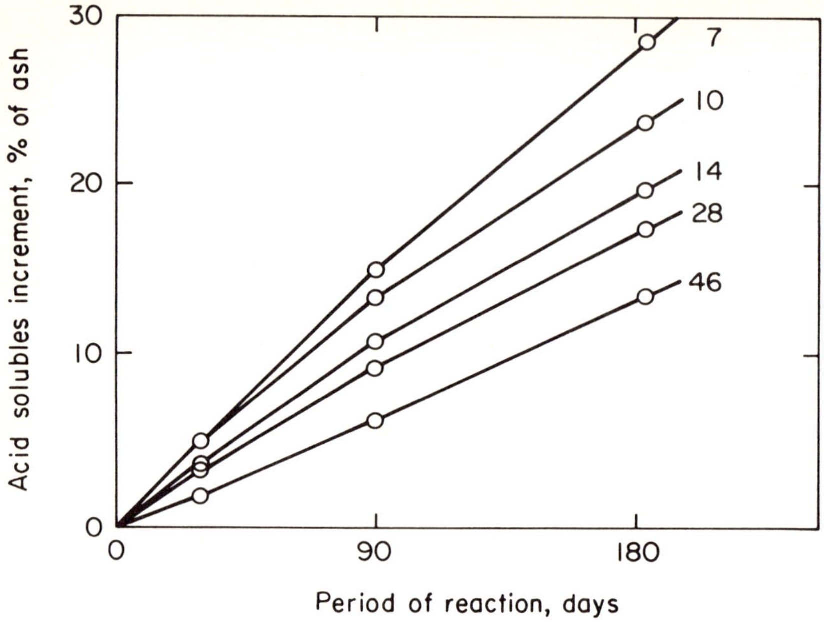

Fig. 9.2 Reaction rates determined by the Feret-Florentin method for fly ashes of various median particle diameters (as indicated, in μm). (Ref. 220)

OH content of the glass, while variable, was dependent on the silica content of the glass and did not directly affect the fly ash reactivity. The iron oxidation state and fictive temperatures of the glasses did not strongly affect the fly ash reactivity but were quite similar for all of the fly ashes studied. The long-term (1-year) degree of reaction in limewater increased progressively with the silica-plus-alumina content of the glass, ranging from 28 to 100% reacted as the (SiO_2 + Al_2O_3) contents increased from 68 to 82%.

When reacted with calcium hydroxide in the presence of NaOH and KOH, the pozzolanic reaction rates were about seven times those found with limewater alone. These results were obtained using fly ash to $Ca(OH)_2$ to water ratios of 2:1:30. In such dilute suspensions the alkali released from the reacted glass would form only very dilute solutions; thus, the reaction in limewater would be much slower than in a dense paste.

Another point to be kept in mind is that the degree of reaction was determined by measuring the IR absorption at 3450 cm^{-1} due to the 0-H bonds in the hydrated gel. The method was made quantitative by calibration with synthetic C-S-H gels substituted with Al and Fe. Unfortunately, only the OH contents were so measured; no attempt was made to determine the Ca to Si ratios of the gels. Since it is known that the H_2O to Si ratio varies with the Ca to Si ratio of such hydrates,[(209)] there may be considerable uncertainty in the extent of reaction if the Ca to Si ratios varied.

Pozzolanic Activity and Strength

In the majority of studies of the pozzolanic properties of fly ashes, the extent of the chemical reactions is not measured at all. Instead, the pozzolanic activity is assessed by tests of strength development during reaction with lime or portland cement. In a few studies both the extent of reaction and the strengths have been determined and related to certain characteristics of the fly ashes.

Pozzolanic Efficiency

Verhasselt(203) attempted to relate the pozzolanic quality of seven fly ashes and three other pozzolans to certain physical characteristics and to their strength development. Density, Blaine specific surface, the amount passing the 45-micron sieve, and loss on ignition were used to develop a classification system in an attempt to predict the behavior of these materials. Reactivity was determined by the amount of CaO reacted. Increases in compressive and tensile strength were used to determine pozzolanic activity. The ratio of strength increase to the increase in CaO that reacted was defined as the pozzolanic efficiency. Pozzolanic efficiency was compared to that of the aluminum hydroxide reaction with CaO, and the results were used to define an index of the quality of the pozzolan. This index was then used to classify materials for different uses, including that as an admixture for cement. However, no systematic relationships between the pozzolanic quantities and the chosen characteristics (which did not include important compositional variables) were found for these diverse materials.

Thorne and Watt's Mortar Tests

Thorne and Watt(123,219,220) used the same 14 well-characterized fly ashes referred to above to identify the most important physical and chemical characteristics that determine the pozzolanic properties of fly ashes as indicated by strength tests. Six "synthetic fly ashes" of controlled compositions were also included in this study. Mortars made with fly ash, sand, and CaO in the ratios 2.5:2:1 were mixed with limited water and compressed into small cubes at 1400 psi (10 MPa) to standardize the degree of compaction and the final water content of the test specimens. These were cured either (1) at 20° C in moist air for periods of up to 730 days, or (2) under the accelerated conditions recommended by Lea: 7 days in moist air at 20° C, 94 hours at 50° C, and 2 hours at 20° C. The proportions used in these mixes were those found to yield maximum strengths. The very low porosities resulted in ultimate strengths (as high as 13,800 psi, or 95 MPa) that were comparable with portland cement mortars similarly prepared. Although the porosities and lime-silica ratios may be much smaller than in ordinary concretes, the main conclusions reached are consistent with other works. These conclusions may be summarized as follows:

1. Different fly ashes produced mortars that developed strengths at different rates but not in a systematic way, so that the ranking of ashes by strength after short curing periods was not the same as at later ages.

2. The effect of curing temperature differences on strength development also yielded variable results. Strengths of mortars cured under accelerated conditions varied from 55 to 175% of the 91-day normal cure strength for different fly ashes.

   The strengths of the accelerated cure (at 50° C) mortars increased linearly with the degree of reaction (7 to 17%) indicated by the Feret-Florentin acid soluble increments. The strengths of the mortars cured at 20° C also correlated well with the degree of reaction up to 182 days, but at 365 days showed no clear correlation with the degree of reaction, which then ranged from 18 to 24%. The strengths also correlated well with the degree of reaction at 20° C as determined by the Lime Solution Method up to 182 days. The Lime Solution Method results did not correlate well with the strengths of mortars cured under accelerated conditions.

   Supplementary tests at various curing temperatures up to 60° C showed that the logarithm of the initial rates of increase of strength were linearly related to the reciprocal of the absolute temperatures. A comparison of the rates of strength development at 20° C and 50° C for three of the fly ashes is shown in Fig. 9.3. Although the initial rate of strength development was increased by a factor of 10 at the higher temperature and the maximum strength was reached in about 40 days, the maximum strengths attained at 50° C were much lower than those attained at 20° C at later ages.

3. Crushing strengths up to 91 days, or under accelerated curing conditions, were correlated only loosely, or not at all, with carbon content, silica and alumina content, glass content, density, and the specific surface of the ashes as single variables. The best correlation was with specific surface as determined by particle size analysis, implying that at early ages the amount of reaction is determined mainly by the surface area of the fly ash particles.

   This conclusion was confirmed in similar tests with fly-ash-like glasses of controlled compositions and finenesses in mortars subjected to accelerated cure. Strengths increased rapidly with increasing fineness from 50 to 100 m^2/kg but less rapidly at higher finenesses; above 300 m^2/kg strengths increased very little. This decreasing effectiveness of increasing fineness was not observed (in this range) with actual fly ashes and was suspected to result from differences in particle size distributions between actual fly ashes and the synthetic preparations. When a factor for the width of the particle size distribution was used with the specific surface area, a good linear relationship with strength was obtained. The substantial differences in chemical composition of these glassy particles had a negligible effect on strengths developed during the accelerated cure.

4. After curing times of 365 days or more, the strength correlated most closely with the SiO_2 or (SiO_2 + Al_2O_3) content of the ash. This finding was interpreted to mean that the long-term pozzolanic activity depended mainly on the amountof material present that was able to react with lime, and not that the reactivity the glass varied with its composition.

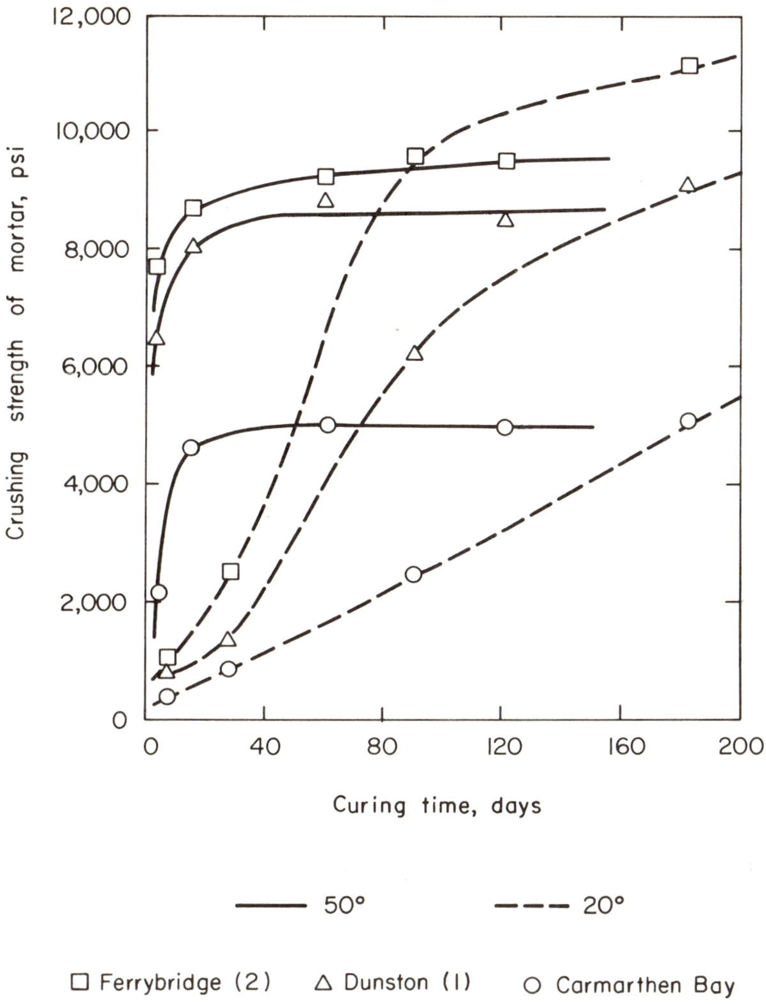

1000 psi = 6.894 MPa

Fig. 9.3 Compressive strengths of lime-fly ash mortars during curing at 20° and 50° C. (Ref. 219)

5. The very high-density (above 2.74 g/cm^3) fractions of fly ashes contained high percentages of particles of high-iron content, showed low reactivity by the Feret-Florentin test, and gave low strengths. The nonsiliceous components of the high-iron fraction, mainly magnetite and hematite, apparently had no pozzolanic activity. Three lower-density fractions contributed equally to strengths on the basis of the weights of the fraction per unit volume of mortar, but the spongy glass particles yielded mortars of much lower strengths because of their low densities (below 2.1 g/cm^3). It was concluded that the pozzolanic activity resides mainly in the glassy siliceous particles.

6. Extraction of the fly ash samples with water before mixing the mortars produced mortars with strengths 2 to 3 times as high as mortars made with unwashed ashes, as shown in Fig. 9.4. The increased strengths were attributed to the dispersion of agglomerates, increased specific surface, and improved homogeneity of the ashes.

Synthetic Fly Ashes of Controlled Compositions

Joshi and Rosauer[204] also attempted to separate the effects of variations of chemical composition, degree of crystallinity, and specific surface area on the pozzolanic activity of fly ashes. Following Watt and Thorne, they prepared synthetic "fly ashes" by a thermal spheroidization process using mixtures of pure kaolinite, calcite, and pyrite so that samples of different crystallinity and particle size could be prepared at several constant compositions. Small specimens of fly ash-$Ca(OH)_2$ mixtures in a weight ratio of 2:1 were mixed with sufficient water for maximum density; compressed specimens were moist-cured at 23° C in an N_2 atmosphere to prevent carbonation. Compressive strengths were determined after curing for periods of up to 28 weeks.

Characteristics of some of the fly ashes prepared from pure kaolinite are given in Table 9.1. The K-Cyc is a synthetic ash prepared from kaolinite and collected in a cyclone separator, and K-Furn is ash from the same preparation which was deposited in and collected from the furnace. K-Recrys is the K-Cyc ash after heat treatment and slow cooling to induce crystallization.

Reexamination of their results leads to the following conclusions:

1. Strength and rate of strength development increased with specific surface area of the K-Cyc ash from 180 to 296 m^2/kg for curing periods of up to 8 weeks, but a further increase of fineness to 518 m^2/kg did not produce a proportional increase in strengths. At early age (10 days), the strength with the 518 m^2/kg ash was actually less than at 296 m^2/kg. These results were in accord with those of Thorne and Watt and were interpreted in terms of the higher water requirement and increased porosity of the mixes made with the finest ash.

2. The strengths increased approximately linearly with a curing time of up to 8 to 10 weeks, as shown in Fig. 9.5. At later ages the strengths increased linearly with the square root of the curing time. It was concluded that the pozzolanic reaction was initially reaction controlled and after 8 to 10 weeks became diffusion controlled.

3. Joshi and Rosauer concluded from the results shown in Fig. 9.5 and the values given in Table 9.1 that the amount of glassy phase in different fly ashes of the same composition influences pozzolanic strength to a greater extent than does particle size. However, they also questioned the validity of their reported values of glass content (see Table 9.1, note **). Since other workers have reported that 1% HF dissolves the glass efficiently, these glass content values are undoubtedly too high.

4. The presence of iron in an otherwise pure aluminosiliceous ash reduced strength development substantially [less than 1600 psi (11 MPa) compared with

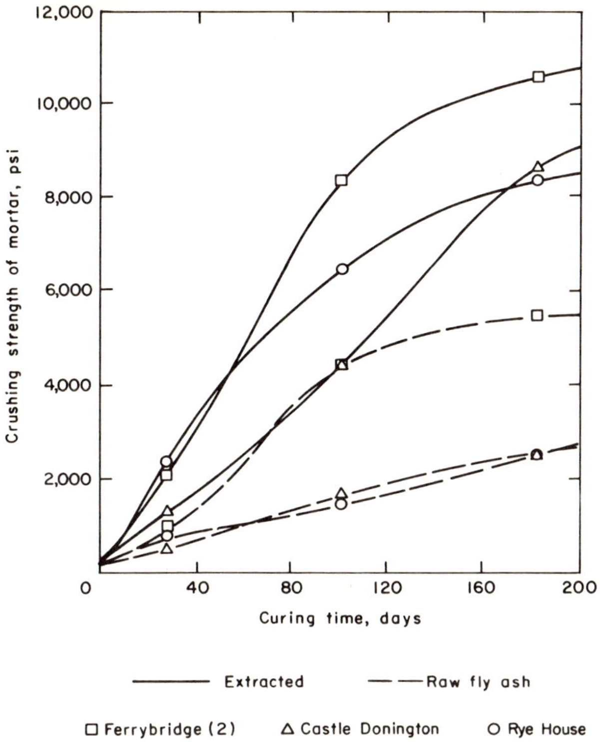

1000 psi = 6.894 MPa

Fig. 9.4 Compressive strengths of lime-fly ash mortars made from raw fly ashes and from fly ashes exhaustively extracted with water before use. (Ref. 219)

Table 9.1 Characteristics of Some Synthetic Fly Ashes* Prepared by Joshi and Rosauer[204]

Fly Ash	Particle Size Fraction (μm)	Blaine Specific Surface (m^2/kg)	Glass Content** (wt. %)
K-Cyc	<5	518	89
K-Cyc	5-10	296	86
K-Cyc	10-20	254	86
K-Cyc	>20	183	87
K-Furn	5-10	—	77
K-Furn	10-20	—	
K-Furn	20	—	
K-Recrys	(whole ash)	—	41 (0?)

*Approximate composition: 52% SiO_2, 46% Al_2O_3, 0.3% CaO, 0.4% Fe_2O_3.
**The authors stated that these values, which were obtained from the amount dissolved in ice-cold 40%HF, are probably too high; XRD revealed no glassy material in the K-Recrys sample, yet the HF method indicates that 41% is glassy.

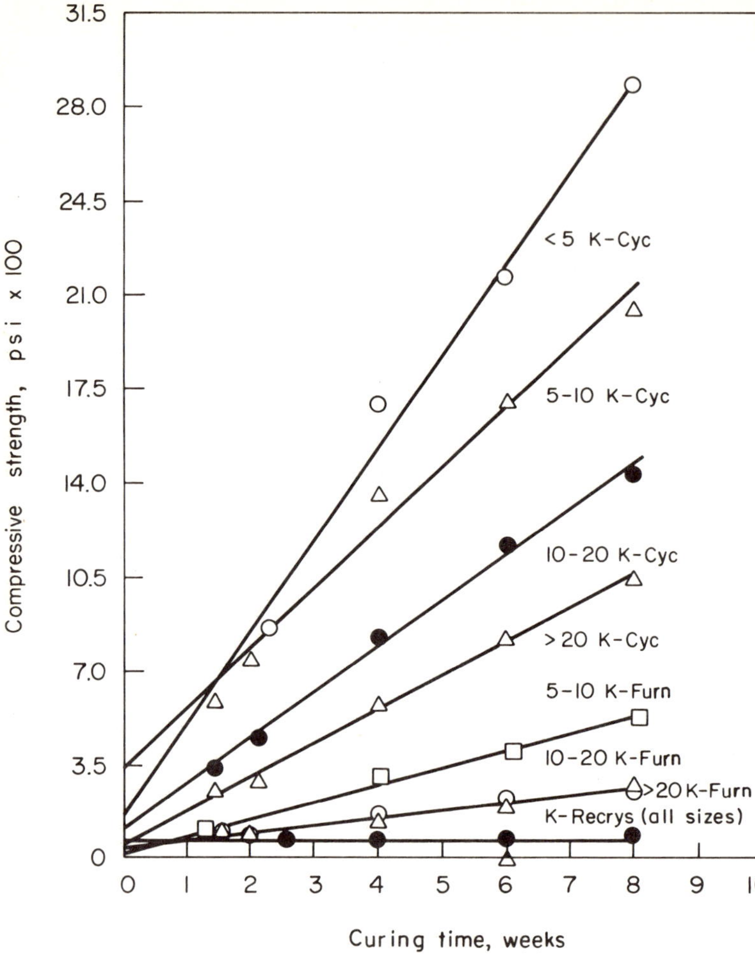

Fig. 9.5 Effects of degree of crystallinity and particle size of a synthetic kaolinite ash on development of compressive strength. (Ref. 204)

2400 psi (17 MPa) at 8 weeks], even when all but about 1% Fe_2O_3 was removed from the whole sample by magnetic separation. Since the glass content (as determined) was not significantly reduced by iron additions, the iron had apparently made the glassy phase less reactive.

5. Fly ashes made with calcium additions had increased specific surface areas, and they produced specimens with improved strengths. The improved strengths were thought to be related to the increased specific surface areas and possibly the presence of hydraulic cement compounds.

6. Calcium in the three-component fly ashes offsets the deleterious effect of iron. These ashes also contained large numbers of fine particles and had relatively high specific surface areas.

Pulverized Coal Fly Ash Specific Surface Effects

Cabrera and Gray[145] attempted to correlate the pozzolanic activity of nine different samples of whole pulverized coal ashes with their chemical compositions and specific surface areas. They determined specific surface areas from particle size distributions, air permeability (Fisher Model 95), and nitrogen absorption on dried samples and also on samples burned at 700° C for 6 hours to remove unburned coal residues. Results of the specific surface area determinations were given in Table 4.1. It was also shown that unburned carbon correlated positively with the specific surface areas by both air permeability and nitrogen absorption on the dried samples.

Compacted mortar specimens were prepared for compressive strength tests from mixtures of fly ash, silica sand, and hydrated lime (94.3% calcium hydroxide) in the proportions 9:40:1, respectively, and a moisture content of 10% for a water to (fly ash + lime) ratio of 0.5. Single variable correlation coefficients of strengths with specific surface areas were very poor. They were, in fact, negative for all four methods at all three ages of test (7, 28, and 56 days) except for the specific surface calculated from the particle size distribution and 56-day strengths. The negative correlations evidently result from the deleterious effects of increasing carbon contents. This study also illustrates how the effect of the specific surface of the reactive glassy phase can be completely masked and how important the unburned carbon content can be. It should be noted that two of these ashes contained exceptionally high carbon contents (25.4 and 53.9%) and that the proportions used for the test specimens were exceedingly lime deficient.

By using the air permeability method, Ravina[147] recently correlated the pozzolanic activity of fly ashes with their finenesses as determined by the amounts retained on a 45-μm sieve and also by the specific surface areas. As described in "Fineness" in Chapter 4, eight low-carbon ash samples were collected from different sections of the electrostatic precipitator in a power station burning bituminous coal. The chemical compositions of the ash samples varied only slightly with fineness. The pozzolanic activity index with lime and also with portland cement was determined according to ASTM C311. Both tests yielded excellent linear functions of the specific surface areas, as shown in Figs. 9.6 and 9.7. Figure 9.7 shows that for specific surface areas in excess of 450 m^2/kg, the fly ashes yielded higher strengths than the replaced cement. The correlations of strengths with amounts retained on the 45-μm sieve were not as good. Ravina recommended that the air permeability method be reintroduced into ASTM Standard Specification C618, from which it had been removed, subject to a limitation on carbon content.

The excellent correlation found by Ravina evidently does not hold when fly ashes from different sources are compared, as shown in the study of fly ashes from 14 power stations by Weshe and vom Berg.[148] They found that early strengths correlated poorly with specific surface areas and that a somewhat

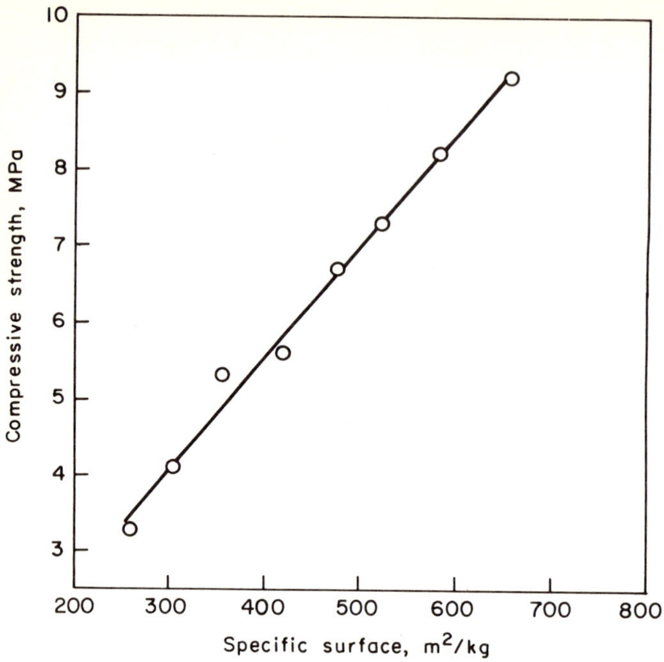

1 MPa = 145 psi

Fig. 9.6 Correlation between specific surface of fly ashes and the Pozzolanic Activity Index with Lime. (Ref. 147)

better correlation was obtained with the amounts finer than 40 μm, especially at later (90 days) age.

In another study Ravina[227] showed that coarse fractions of fly ashes (260 and 304 m^2/kg) contributed significantly to strength in precasting concretes when cured at 75° C.

Steam Curing of Plastic Mortars

Valore investigated methods of evaluating fly ashes and other pozzolans for use in the production of steam-cured and autoclaved concrete products.[228] For such purposes he found tests of plastic mortars of uniform flow table consistency to be most suitable. Two types of mortars were used. Mortars made with ASTM C 109 standard graded sand were used for studying various potential combinations of binding materials. Mortars made with minus No. 30 sieve fractions of aggregates proposed for use were used in the same binder-aggregate ratios as in the concrete products. In this way potentially reactive fines in the aggregates are included with the binder as in plant-made products. Formulations of binders containing portland cement, raw fly ash, beneficiated fly ash, burned shale fines, and silica flour were tested using various proportions of portland cement. Data for six well-characterized fly ashes are given.

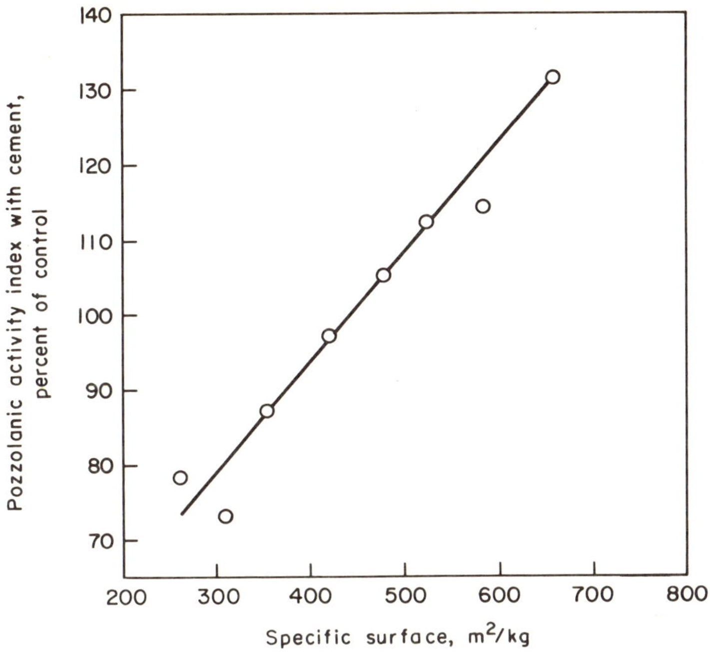

Fig. 9.7 Correlation between specific surface of fly ashes and the Pozzolanic Activity Index with Portland Cement. (Ref. 147)

Tests of strengths at various ages of mortar cubes at both ordinary temperatures and after steam curing and autoclaving often show pronounced maxima at particular pozzolan percentages (typically 40% pozzolan). These effects are more pronounced in the mortars than in concretes. It is also shown that pozzolan additions are of doubtful (or negative) value when added to mixes in which the aggregate fines have pozzolanic value.

Up to this point we have reviewed the effects of fly ash's physical characteristics, compositions, and the reaction environment on reactivity, microstructural development, and strength in cementitious mixtures. We have seen that there are some complex interrelationships between the fly ash characteristics and strength development. Successful prediction of concrete properties from such data on fly ashes meeting current specifications will apparently require more detailed modeling of these processes than has yet been done. Another approach has been to measure the reactivity directly by chemical methods to separate reactivity from other characteristics that can also influence concrete properties but perhaps not in the same ways as in the test methods. Some of these chemical methods are reviewed in the next chapter.

Chapter 10
Chemical Methods to Assess Reactivity

Definitions

Many investigators have attempted to devise simple chemical methods to evaluate fly ashes for their pozzolanic properties. Such methods have been reviewed by Malquori,[199] Kokubu,[3] Massazza,[201] Clifton et al.,[186] and Sersale.[127] Methods that would be much less time consuming than pozzolanic activity tests based on strength development would be extremely useful. No such method has yet been definitely established, but several methods have shown some potential utility.

The traditional chemical methods are variations of measurements of the amount of lime that reacts with the pozzolan when mixed with lime or portland cement. Closely related to such tests of the pozzolans are tests of the pozzolanicity of a cement-pozzolan blend. Pozzolanic cements, in the European usage of the term, are blends of portland cements and pozzolans that contain sufficient pozzolan to combine (or immobilize) all of the calcium hydroxide resulting from hydration of the clinker. Cements containing only small amounts of pozzolan are designated "Portland pozzolana cements" by ISO.[187] (Confusion may result from the fact that ASTM C595 designates Type IP and P "Portland-pozzolan cements" as cements containing 15 to 40% blended ingredients.)The pozzolanicity test was conceived to determine whether a cement can be considered to be truly pozzolanic without knowledge of the composition of the portland cement, the proportions used, or other properties of the pozzolan.

Pozzolanicity Test

Description

The pozzolanicity test proposed by Fratini[229] is intended to determine the ability of the blended cement to produce hardened pastes practically devoid of calcium hydroxide. The test is part of the cement specifications in France, Italy, Spain, and Portugal and accepted by ISO.[230] The test consists of storing a mixture of 20 g blended cement and 100 mL distilled water in a sealed bottle at 40° C for 8 days. The contact solution is analyzed to determine the total alkalinity (OH^-) and the calcium ion concentration. Since the concentration of alkalies strongly affects the calcium ion concentration, the total alkalinity is necessary to determine whether the solution is or is not saturated with respect to calcium hydroxide. A lime saturation curve (see Fig. 10.1) is used for this purpose. If the representative point falls below the curve, the pozzolanicity test result is positive; if the point is above the curve, the result is negative; if the result is in doubt, the test is repeated and continued to 15 days.

Critical Evaluation

The method has been critically evaluated by Luxan and Soria[231] using four portland cements (ordinary and 0% C_3A, sulfate resistant), three pozzolans, and pure quartz as a control. Analyses were performed at 7 and 28 days; the results at 7 days for the intermediate alkali content (2.73% Na_2O and 1.72% K_2O) pozzolan used at five different addition levels are shown in Fig. 10.1. At addition levels below 20% pozzolan, the solutions were all supersaturated with respect to calcium hydroxide, a condition characteristic of portland cements. It is well known that very small crystals are more soluble than the same crystals in bulk form. At 30% pozzolan or more, the solutions were unsaturated with respect to calcium hydroxide, i.e., these cements were pozzolanic.

Luxan and Soria found that it was very difficult to avoid carbonation of the calcium during the test. In addition, as indicated in Fig. 10.1, the results at 7 days with increasing pozzolan percentages did not always produce systematic reductions in calcium concentrations. At 28 days these reductions did become more systematic. They concluded that the method did not discriminate with great precision, although it was useful for differentiating inert from active additions. It has been used to determine the minimum quantity of fly ash necessary to prepare a pozzolanic cement.[232]

Proposed Modification

Bonomi and Francardi[233] have proposed a simple modification of the pozzolanicity test which avoids (1) the possible interference of other ions (Sr^{2+} and SO_4^{2-}) in the determination of the calcium concentration, (2) the need to determine the total alkalinity (or Na^+ and K^+ concentration), and (3) the use of the lime solubility curve. Instead, the test solution is analyzed as before to obtain the calcium oxide concentration $[CaO]_A$. Then a sample of this same solution is mixed with an excess of finely ground $Ca(OH)_2$ to produce a solution saturated with respect to solid calcium hydroxide, which is also analyzed to determine $[CaO]_B$. If $[CaO]_B$ is greater than $[CaO]_A$, the contact solution is unsaturated and the cement is pozzolanic. If the reverse is true, the contact solution is supersaturated, indicating that the cement is not truly pozzolanic.

Significance of Results

The fact that the solution becomes unsaturated with respect to solid calcium hydroxide does not necessarily mean that all of the calcium hydroxide has been reacted. Examination of pozzolanic cement pastes by XRD, DTA, and by extraction of pulverized material all show the presence of some residual calcium hydroxide.[199] The presence of this residue is explained by the formation of protective coatings of reaction products on the calcium hydroxide crystals, which impedes its passage into solution. In this way a portion of the calcium hydroxide becomes effectively immobilized. This process must be considered in any evaluation of pozzolanic activity and also with respect to the structure and properties of the hardened cements. Some pozzolanic activity tests avoid encapsulation of calcium hydroxide crystals by using relatively large volumes of

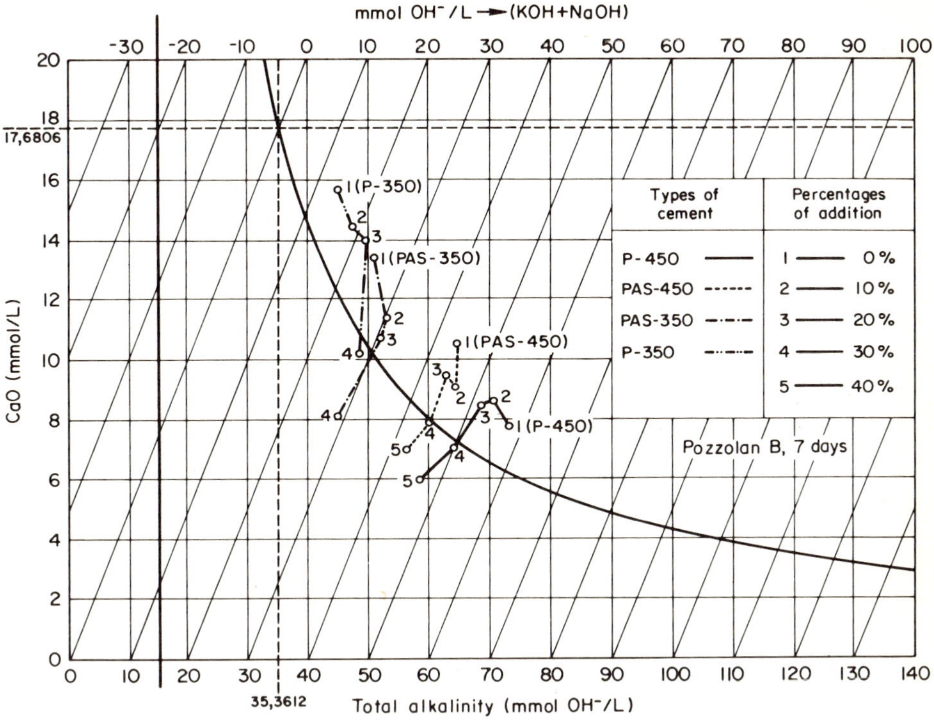

Fig. 10.1 Pozzolanicity tests of four different portland cements with a particular natural pozzolan at five different cement replacement levels. (Ref. 231)

lime-saturated solutions as the source of lime.

Bensted[234] pointed out that if a cement containing a highly reactive alite is used in the pozzolanicity test, a negative result may be obtained with a material which has pozzolanic properties. He also indicated that the use of a sulfate resistant cement would not be suitable because the consumption of lime by the hydration of the ferrite phase reduces the pozzolanic reaction appreciably. Sersale[127] has objected that the use of the pozzolanicity test for characterizing the activity of pozzolans is improper. However, the test does provide information concerning the relative rates of reactions at early ages of blends of portland cement and pozzolans.

Chemical Tests of Fly Ash Reactivity

Lime Consumption Tests

The oldest test for pozzolans is that of Vicat, in which the rate of absorption of lime from a calcium hydroxide solution is measured. The traditional chemical method for measuring the reactivity of a pozzolan is the determination of the amount of lime which combines with the pozzolan when mixed with hydrated lime or portland cement. The amount of uncombined lime can be obtained by (1) extraction with suitable solvents (e.g., aceto-acetic ester-isobutanol), (2) calorimetry, (3) DTA-TGA, or (4) extraction with partially saturated lime solutions.[7B] The combined lime can then be determined by difference. In the case of mixtures with portland cements, the amount of lime released by hydration of the cement must be known or estimated. This requirement may present difficulties because, as we have already seen, fly ash can affect the rate of hydration of the cement.

Comparison of Four Methods

Watt and Thorne[220] reviewed a number of chemical methods in connection with their studies of fly ash activity and selected four as being representative of most of the available methods. It should be noted that these four methods are used to predict reactivity, whereas the Feret-Florentin method used by Watt and Thorne with mortars was used merely to determine the extent of the lime-fly ash reaction in mortars. Results obtained by the four methods were evaluated for reproducibility and for the relationship to mortar strength:

1. Baire's Method (Sestini and Santarelli[235] modification): The fly ash was extracted with HCl and then with KOH and analyzed for SiO_2 + Al_2O_3 + Fe_2O_3. Determinations on four ashes yielded 3 to 7% of the oxides dissolved in HCl and 19 to 26% in KOH. The reproducibility was poor, and there was no apparent relationship with the strengths of the mortars.
2. Lea's Conductometric Method:[236] The fly ash was allowed to react for a few hours in a closed cell with a calcium hydroxide solution; the conductivity of the solution was measured periodically to follow the reaction of lime in tests of 14 ashes. The amount of reaction was small and reproducibility was poor. The results showed no relationship with the strengths of mortars.
3. Moran and Gilliland's Method:[237] The fly ash was allowed to react with saturated limewater in a sealed tube; the solution was back-titrated to determine how much lime had been reacted. Two different sample weights, 0.2 and 0.5 g, were used with 50 mL of solution for periods of 7, 28, and 91 days. The method showed good reproducibility and discrimination, with results for different ashes spread over a wide range. However, the results with the different sample sizes did not agree and also showed no agreement with the strengths obtained in Watt and Thorne's tests. However, in their tests Moran and Gilliland had found that the results could be related to mortar strengths, perhaps because the sample-solution ratios were higher than in Watt and Thorne's tests.
4. The Feret-Florentin Method and the Lime Solution Method: These have already been described and the results given in "Fineness" in Chapter 9. These methods were not used to predict the long-term behavior of an ash in the mortar as much as to provide a convenient means of measuring the amount of

pozzolanic reaction that was taking place during these tests. The reproducibility of both methods was found to be good. However, the results for both methods at 50°C showed only a loose correlation with the long-term strengths of mortars at 20°C. Watt and Thorne concluded that these methods have no prospective value for assessment of long-term strength development.

The finding that Baire's and Lea's methods showed poor reproducibility and that Lea's method involved only a very small amount of reaction would seem to eliminate them from further consideration. Some variation of Baire's method, however, may warrant further consideration. The initial extraction with HCl was found useful in the Feret-Florentin method to remove nonpozzolanic acid soluble material (such as hematite). Acid extraction would also remove cementitious phases; these could probably be selectively extracted before the HCl extraction. Finally, the glassy material could be extracted with alkali or HF. The sum of the cementitious and glassy phases could be considered to be the strength-producing active material.

The validity of Watt and Thorne's conclusions with respect to the utility of these chemical methods depends upon certain assumptions. It was tacitly assumed that the fly ash-lime mortar strengths were a suitable measure of the pozzolanic activity that these fly ashes would show when used in portland cement concrete. There are at least two reasons to doubt that these mortar tests were suitable.

First, there is the matter of the mix proportions. As discussed earlier, these mixtures have very low lime-silica ratios. The very low water contents and compaction procedure yielded very dense specimens which achieved very high strengths at relatively low degrees of reaction. Because of the very low initial porosities, these test conditions do not at all simulate the situation that prevails when fly ashes are used in plastic concrete mixtures. However, the low degrees of reaction reached because of the low porosities tend to compensate for the low lime-silica ratios of the mixtures. In addition, the low water contents mean that the mixtures have high alkali concentrations, which would increase the reaction rate.

Second, there is the difference between reaction with lime and with portland cement. Alkalies released from the cement also accelerate the reaction. As was shown earlier, portland cements not only provide calcium hydroxide for reaction, but reaction also occurs with the calcium silicate hydrate produced by cement hydration.

For these reasons we cannot be certain that the lack of correlation of the chemical methods with Watt and Thorne's mortar strengths would imply a lack of correlation with strength development in tests that more closely simulate the conditions in concretes.

Chapelle Test

An accelerated lime consumption test has recently been modified and evaluated by Raverdy et al.[238] In this Chapelle Test, a sample of pozzolan is reacted with lime in water at 100°C for 16 hours. The method was improved with better temperature control and magnetic stirring.

In tests of various pozzolans, including one fly ash, the method did not correlate well with the percentage of glass. Since the natural pozzolans contained devitrified reactive material, these results may not be indicative of the utility of the method with fly ashes; further investigation is required.

Proposed Conductivity Method

Raask and Bhaskar[239,240] developed a rapid method to determine the pozzolanic activity of fly ashes by measurement of the increase of conductivity of a 0.1 molar (2%) HF solution as the fly ash dissolves in the acid. They attributed the increase in conductivity to dissociated fluosilicic acid. Although the changes in conductivity may not have been an accurate measure of the silica concentrations in solution, the rate of solution data obtained yielded good fits to the rate equation for diffusion-controlled reaction. The data also correlated very well with strength data expressed in terms of cementing efficiency factor, *K* ("Proportioning of Concrete with Fly Ash" in Chapter 1). The absolute magnitudes of the silica concentrations determined from the conductivity were questionable because the method was calibrated using nearly pure silica fume, and it has been shown that Fe_2O_3, TiO_2, Al_2O_3, CaO, MgO, Na_2O, and K_2O are also extracted in appreciable amounts from the fly ash.[135,241] The method may be useful even if the conductivity is affected by other constituents, because the indicated rate of solution (of the glassy phases) is directly related to the pozzolanic activity, as indicated by the cementing efficiency.

Both processes are probably governed by similar diffusion-controlled dissolution processes. The rates, however, are very different; the amount of fly ash dissolved in HF after 1 hour at room temperature is approximately the same as the amount dissolved (i.e., reacted) in the alkaline media of hydrating portland cement after 1 year.

The initial rate of solution in HF was found to fit the equation

$$w_r = w_0(1 - \beta e^{-\alpha t}) \qquad (10.1)$$

where w_r is the amount reacted and w_0 is the total silica. The rate constant, α, is related in diffusion theory to the diffusion coefficient, *D*, the density, *d*, and the specific surface area, *S*, according to

$$\alpha = \frac{\pi^2 D d^2 S^2}{9} \qquad (10.2)$$

The fact that the rate constant is proportional to the square of the specific surface area was experimentally verified by measurements with samples of the lowest carbon content ash sample used. This ash was ground to various finenesses up to 3900 m^2/kg as determined by nitrogen adsorption. The results yielded a good linear relationship between α and S^2.

It should be noted that Eq. 10.2 could be verified only because the amount of carbon residue was the same in each sample and because the nitrogen surface area of the carbon apparently did not change during grinding. With different ashes ranging in carbon content from 0.7 to 11.1%, the relationship did not apply.

Values of α for 12 fly ashes were obtained from the experimental data using Eq. 10.1. For convenience, a quantity called the pozzolanic index (P_z) was defined as

$$P_z = 1000\alpha \qquad (10.3)$$

Values of P_z ranged from 0.40 to $1.25 s^{-1}$ for these 12 fly ashes. An approximately linear relationship was found between this pozzolanic index and the pozzolanic activity as indicated by the cementing efficiency shown in Fig. 10.2. These values were used to classify fly ashes according to their relative merit as materials for use in concrete, as given in Table 10.1. Although there are a number of uncertainties concerning the relationships between the pozzolanic index as determined by this method and the strength results, the method does appear to warrant further investigation as a very rapid and accurate method for characterizing fly ash for pozzolanic reactivity and strength development. The method is directly sensitive to two of the most important fly ash characteristics, the glass content and fineness. Its sensitivity to the third characteristic (SiO_2 + Al_2O_3) is not yet clear, nor is there enough information concerning other factors that may interfere with the accuracy of the method.

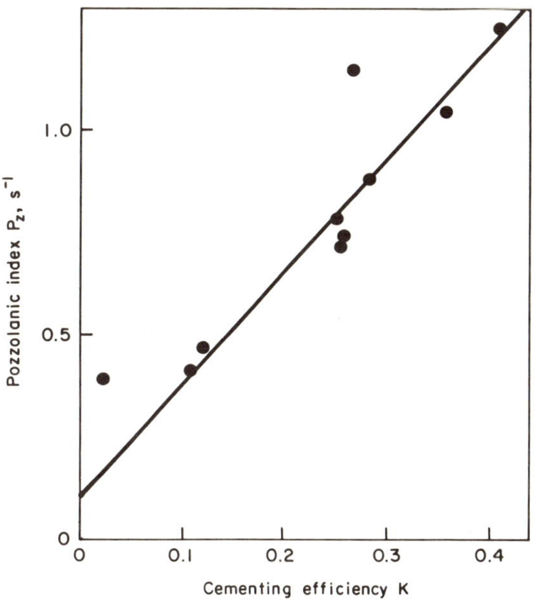

Fig. 10.2 Relationship between pozzolanic index (P_z) and cementing efficiency (K). (Ref. 239)

Table 10.1 Classification of Ashes by Pozzolanic Index, P_z[239]

Classification	Poor	Satisfactory	Excellent
Pozzolanic index	0.0 to 0.5	0.5 to 1.0	1.0
Cementing efficiency (K)	0.0 to 0.15	0.15 to 0.33	0.33

Calcium Oxide Heat Evolution Test

In the section "Standard Specifications and Test Methods" in Chapter 1 and the section "Variability of Fly Ash Properties" in Chapter 4, a simple test used to measure the calcium oxide content of Class C fly ashes was described as the "Acid Temperature Rise" test.[80,81] This test has been found to be useful as a rapid check for uniformity of cementitious fly ashes.

Chapter 11
Durability Considerations

Soundness

The effects of fly ash on the soundness of cements and concretes have been considered mainly in terms of excessive free CaO and MgO contents. It is generally believed that the autoclave expansion test provides adequate information to protect against unsoundness caused by the use of fly ash, but there does not appear to be much supporting data. Even in the case of portland cements without fly ash, there is some question whether the autoclave expansion test (ASTM C 151), or any accelerated test, may be too restrictive.[(242)] Free CaO is a recognized cause of unsoundness, especially in low-water content mixtures.[(75)] Cain[(89)] reported that some Class C fly ashes fail the autoclave test but do not cause unsoundness in concretes; he explained that the wetter mixes used in concretes permit more complete CaO hydration before final set than do the drier autoclave test mixes. One engineer's experience with numerous fly ash sources in the midwestern United States has shown that all failures of the autoclave test attributed to free CaO occurred with Class F and not with Class C ashes.[(243a)] Expansion in the test, and in hardened concrete, depends not only on the free CaO content but also on the particle sizes and encapsulation in the glass. Class C fly ashes are often very fine, and the free CaO may hydrate harmlessly (except for heat effects) before the concrete hardens. Encapsulation of CaO in the glass, as reported by Adams,[(128)] may be more likely in coarser fly ashes. Such inclusions would react only after water has penetrated the glass coating.

On the other hand, a recent study by Schlorholtz and Demirel[(243b)] of the soundness of cements made with one low CaO content (2%) fly ash and four high CaO content (14 to 30%) fly ashes showed large (1.5 to 5.8%) autoclave expansions. The expansions above critical fly ash contents of 20%, 25%, and 36% for three of the fly ashes were very sensitive to fly ash content, increasing abruptly at these critical percentages for these fly ashes. ASTM C311 requires that the autoclave test be done using a fly ash content of 20% by weight, unless higher fly ash contents are used in project mix designs; in those cases the test specimens shall contain the project mix design percentage. However, these authors note that the high sensitivity to fly ash content at the critical percentage may require quality control that onsite batch plants may not be able to maintain. They also provided a thermodynamic calculation of the maximum stresses that may be required to completely restrain expansion by hydration of CaO and MgO, 5×10^5 psi (3500 MPa) and 3×10^5 (2000 MPa), respectively, far in excess of the strength of the restraining matrix.

For many years MgO contents of fly ashes were limited in ASTM C 618 to less than 5% to guard against the presence of large crystals of periclase, which are known to cause unsoundness. It is now generally recognized that most of the MgO is present in glassy phases,[(124)] and the MgO limit has been deleted.

Use of fly ash as a means of controlling the unsoundness of portland cements of high MgO contents (over 6%) has been studied with favorable results.[(244–250)]

Rosa[250] has discussed the mechanism of expansion control in terms of (1) sealing off of periclase grains in clinker grains, and (2) formation of magnesium silicate hydrates. Also, the pozzolanic reaction decreases the C to S ratio of the calcium silicate hydrates and promotes the formation of a gel with a greater capacity for MgO substitution.[246]

The reactions between the sulfates, lime, and aluminates to produce ettringite, as in Eq. 6.3 and in the next step of the reaction of Eq. 7.5, are potential causes of unsoundness as well as sulfate attack. These are the same reactions used in the sulfoaluminate expansive cements in which the magnitudes of the expansions are controlled by the cement composition, fineness, and curing conditions.[251] Free lime and moisture saturation are known to be essential for maximum expansions, and even slight variations from these conditions greatly reduce the expansions. Since the pozzolanic reactions tend to reduce the free calcium hydroxide and also the permeability, the presence of pozzolanic fly ash might be expected to reduce sulfoaluminate reaction expansions unless lime from the fly ash itself contributes to such unsoundness. Several authors have called attention to the free CaO, C_3A, and $C_4A_3\overline{S}$ contents of the high calcium content fly ashes.[27,62,72,125,126] $C_4A_3\overline{S}$ is the principal expansive component in Type K expansive cement; C_3A is the expansive component in Type S. Very deleterious expansions have been reported for some high-calcium content fly ashes.[252]

The mechanism of the expansion is still a subject of controversy and current research. Richards and Helmuth proposed a mechanism that appeared to make possible quantitative predictions of the magnitude of the expansion under both free and restrained expansion conditions.[253] However, this mechanism has not been examined with respect to the use of fly ash or correlated with autoclave test results. Since the conditions in the autoclave are different from normal curing and since the expansive reactions are sensitive to curing conditions, it appears that further research on these questions is urgently needed before fly ashes can be characterized for soundness of concrete with confidence.

Sulfate Resistance

Very similar considerations apply to the sulfate resistance of concretes made with Class C fly ashes. The chemistry of sulfate attack on concrete has been summarized by Hansen.[59] In the case of sodium sulfate solutions, the reactions produce additional ettringite, as with the reaction of tricalcium aluminate with gypsum in Eq. 6.3, if lime is also available. Sodium sulfate also reacts with the available calcium hydroxide to produce gypsum, according to Eq. 6.7, in addition to the reaction with the aluminates to produce ettringite. Alkalies, sulfate, and aluminate may also be incorporated into the calcium silicate hydrate phase, especially when the C to S ratio is reduced by the pozzolanic reaction. This, and the fact that ettringite formation requires a high calcium concentration in solution, appear to be some of the major factors involved in improved sulfate resistance. The lower lime contents in pozzolanic cements result in lower lime hydrates, e.g., gehlenite hydrate and hydrogarnet; these hydrates should not form ettringite when reacted with sodium sulfate because of lack of lime. In addition, Lea has suggested that the C-S-H deposited by the reaction of the pozzolan may form an impervious coating on the alumina-bearing phases.[7B]

Reactions in solutions containing magnesium sulfate (as in sea water) are given by Mather.[60] Hansen has also pointed out that, in addition to the double decomposition reaction of Eq. 6.7, there will be the reaction

$$Ca(OH)_2 + MgSO_4 + 2H_2O \rightarrow CaSO_4 \cdot 2H_2O + Mg(OH)_2 \qquad (11.1)$$

Because the solubility of $Mg(OH)_2$ is only 9 mg/L (pH = 10.5), this reaction tends to go to completion.

The sequence of reactions proposed by Mather[60] (with minor errors corrected) is

$$2C_4AH_{13} + 3M\overline{S}H_7 \rightarrow \underset{\text{(ettringite)}}{C_3A(C\overline{S})_3H_{32}} + 2CH + AH_3 + 3MH + 7H \qquad (11.2)$$

$$CH + M\overline{S}H_7 \rightarrow C\overline{S}H_2 + MH + 5H \qquad (11.3)$$

The gypsum then reacts with C_3A hydrate:

$$C_4AH_{13} + 3C\overline{S}H_2 + 14H \rightarrow C_3A(C\overline{S})_3H_{32} + CH \qquad (11.4)$$

The magnesium sulfate also reacts with the C-S-H:

$$C_3S_2H_x + 3M\overline{S}H_7 \rightarrow 3C\overline{S}H_2 + 3MH + \underset{\text{(silica gel)}}{2SH_y} + (12 + x - 2y)H \qquad (11.5)$$

In addition, the magnesium hydroxide may react to form the magnesium silicate hydrate, which has no binding power:

$$4MH + SH_y + (4.5 - y)H \rightarrow M_4SH_{8.5} \qquad (11.6)$$

In the absence of free calcium hydroxide, magnesium sulfate may react with the ettringite produced in Eq. 11.2 to form gypsum:

$$C_3A(C\overline{S})_3H_{32} + 3M\overline{S}H_7 \rightarrow 6C\overline{S}H_2 + AH_3 + 3MH + 35H \qquad (11.7)$$

With respect to reactions 11.5 and 11.6, the favorable influence of pozzolans on resistance to magnesium sulfate or sea water has been attributed to increased amounts of silica gel formed when the sulfate acts on the C-S-H.[254] Recent research indicates that pozzolanic reactions reduce the permeability by blocking pores, as discussed in "Permeability and Durability" in this chapter.

Research with good quality Class F fly ashes has adequately demonstrated the potential for improved sulfate resistance imparted by fly ash,[29,30,61] as shown in Fig. 1.5. Realization of that potential depends on proper proportioning of the concrete, adequate curing, and also on the composition of the fly ash. Some high calcium content fly ashes drastically reduce sulfate resistance.[62a,62b,255] Dunstan[62a,62b] proposed a chemical criterion to identify fly ashes that improve the sulfate resistance of concretes. He assumed that fly ashes that have compositions that fall in the mullite field on the equilibrium phase diagram (Fig. 3.6) do not react with sulfates to form ettringite, suggesting that the degree to which the alumina in the glass reacts with sulfate to cause deterioration is directly related to the calcium oxide content. He then assumed the decrease in sulfate resistance to be a function of the calcium oxide content in weight percent that exceeds 5%

(or %CaO – 5). He also suggested, on the basis of work by Kalousek and Benton,[256a] that iron-rich ettringite crystals do not grow to cause expansion, or grow very slowly, so that decreased sulfate resistance could be postulated to be inversely proportional to the iron content in percent by weight. He then expressed the total effect on sulfate resistance in terms of a sulfate resistance factor *R*:

$$R = \frac{\%CaO - 5}{\%\mathrm{Fe_2O_3}} \tag{11.8}$$

Although it is recognized that the oxides are present in both the glassy and the crystalline phases in the fly ash, the total oxides are used directly to calculate *R*. Examination of data obtained with specimens soaked in 10% sodium sulfate solutions, or subjected to wetting and drying with 2% sodium sulfate solutions, indicated that an *R* factor of less than 0.75 yielded greatly improved sulfate resistance. An *R* factor of greater than 3.0 indicated a reduced sulfate resistance. For very severe cyclic conditions or for soaking in magnesium sulfate solutions, these values were reduced by 0.5.

In considering the limitations of the *R* factor in predicting concrete performance, Dunstan recognized the fact that if the iron content of the fly ash is very low, the factor tends to infinity, indicating low sulfate resistance. He suggested that a value of 2% be arbitrarily used for such fly ashes. It is also apparent that such chemical criteria should consider the composition of the cement to be used with the fly ash.

In a later paper Dunstan[256b] included the effects of additional factors to estimate the sulfate resistance of concretes. Cements and fly ashes of given compositions were used in concretes of different proportions and water-cement ratios for tests of different severities of exposure to sulfate attack. A sulfate-resistance factor, F_{SR}, was defined that could be calculated for any concrete composition to predetermine its suitability for use in different sulfate exposures.

For portland cement concretes Dunstan expressed F_{SR} in terms of three factors derived from ASTM specifications and other empirically based relationships.

$$F_{SR} = S\left(\frac{\mathrm{w/c}}{0.52}\right)\left(\frac{\%\mathrm{C_4AF}}{2} + \%\mathrm{C_3A}\right) \tag{11.9}$$

in which

S = severity level, based on the ACI levels given in Table 11.1.[24B]

Dunstan used Eq. 11.9 to show that experience embodied in current ASTM specifications indicates that portland cement concretes with F_{SR} less than about 40 should be resistant to sulfates at severity levels *S* in this range.

For concretes containing fly ashes, Eq. 11.9 was modified. For concretes containing high calcium oxide content fly ashes, sulfate resistance decreases at low fly ash percentages; at higher percentages the trend reverses and the sulfate resistance at high enough fly ash contents may be improved. Low calcium oxide content fly ashes of good quality show no such reversal but show improvement at even small additions.

Table 11.1 Severity of Exposure of Concrete Subject to Sulfate Attack[24B]

Severity Level S	Exposure	Water soluble sulfate, SO_4, in soil, percent	Sulfate, SO_4, in water, ppm
1	None	0	0
2	Mild	0.00-0.10	0-150
3	Moderate	0.10-0.20	150-1500
4	Severe	0.20-2.00	1500-10,000
5	Very severe	Over 2.00	Over 10,000

To represent these effects, Dunstan fitted the dependence of the sulfate resistance on the percentage of fly ash in blended cement concretes to a blended material function of the form

$$K_b = 1 + \sin(R_v) - |\sin(R_v - 0.9\,P_b)| \tag{11.10}$$

This blended material factor, K_b, is a function of the weight percent fly ash in the blend, P_b, and a resistance variable, R_v, which depends on the composition of the fly ash.

Dunstan evaluated R_v from two different test methods used at the U.S. Bureau of Reclamation. For continuous soaking in 10% sodium sulfate solutions he found that

$$R_{vs} = 0.67R^2 \tag{11.11}$$

For cyclic soaking in 2% sodium sulfate solutions at room temperature for 16 hours and drying at 54° C (130° F) for 8 hours he found

$$R_{vc} = 4.5R = 4.5\frac{\%CaO - 5}{\%Fe_2O_3} \tag{11.12}$$

For exposure conditions intermediate to these he suggested

$$R_v = (1 - C_c)0.67R^2 + C_c4.5R \tag{11.13}$$

in which $C_c = 1$ for daily wetting and drying cycles and C_c is less than 1 for less frequent cycling.

To further quantify the severity level, S, in terms of cyclic frequency and sulfate solution concentrations he used

$$S = 1 + C_c + \log C_s \tag{11.14}$$

in which C_s is the greater of

$$C_{s1} = \text{parts per million in groundwater or test solution}$$

or

$$C_{s2} = 50\,Y_m(S_s - 0.08) \tag{11.15}$$

in which

Y_m = percent of year in which adequate moisture is present to induce sulfate attack

S_s = percent soluble sulfates in soil

Using these relationships, Eq. 11.9 is generalized to include fly ash blended cement concretes

$$F_{SR} = S(\frac{w/(c+b)}{0.52})(\frac{\%C_4AF}{2} + \%C_3A)K_b \qquad (11.16)$$

in which w/(c+b) is the water to cement plus fly ash ratio.

Fig. 11.1 shows a plot of F_{SR} versus log (days to 0.5% expansion) for tests of concrete cylinders exposed to three different exposure conditions. The time to failure (0.5% expansion) was in many cases estimated by long extrapolations from earlier age data. The linear regession line has a correlation coefficient of 0.87. Dunstan also showed 50 and 75 percent estimated error lines and that a 50-year time to failure would correspond to an F_{SR} value of 40 using the 50 percent error line.

The scatter of the data in Fig. 11.1 indicates about an order of magnitude range in the predicted time to failure for equal values of F_{SR}. Part of the reason for the large scatter is undoubtedly a result of the use of the 0.5% expansion values. Not only does this require long extrapolations for durable concretes, but such expansions are far in excess of elastic expansions, as will be discussed below in connection with test methods. It is evident that further research is required to refine such an approach and to include effects of curing variations.

The need for test methods for sulfate resistance of cements other than actual long-term exposure of concretes was pointed out in "Effects of Fly Ash on Durability" in Chapter 1. It is generally recognized that one of the most important factors affecting sulfate resistance of concretes (with or without fly ash) is their permeability to water and sulfate solutions. Test methods differ in the degree to which the potentially large protective effect of reduced permeability is recognized.

Three test methods have been used for comparing the sulfate resistance of hydraulic cements in ASTM interlaboratory test programs. ASTM C452, Potential Expansion of Portland Cement Mortars Exposed to Sulfate,[(21B)] is a test using mixtures of portland cement with added gypsum in mortars to determine the potential sulfate resistance of portland cements by expansion measurements. This test avoids the effects of differences in permeability on the rate of sulfate attack, but results with blended cements failed to yield results that correlate with experience with mortars or concretes matured before exposure.[(79b)]

Another test, the Wolochow test,[(257a)] is a lean mortar bar test in which specimens are cured for 7 days before immersion in 5% sodium sulfate solutions; the expansion after 28 days is sufficient to detect portland cements of low sulfate resistance. Tests in different laboratories were not sufficiently reproducible for this test to become an ASTM Standard Test Method, but it was nevertheless widely used.

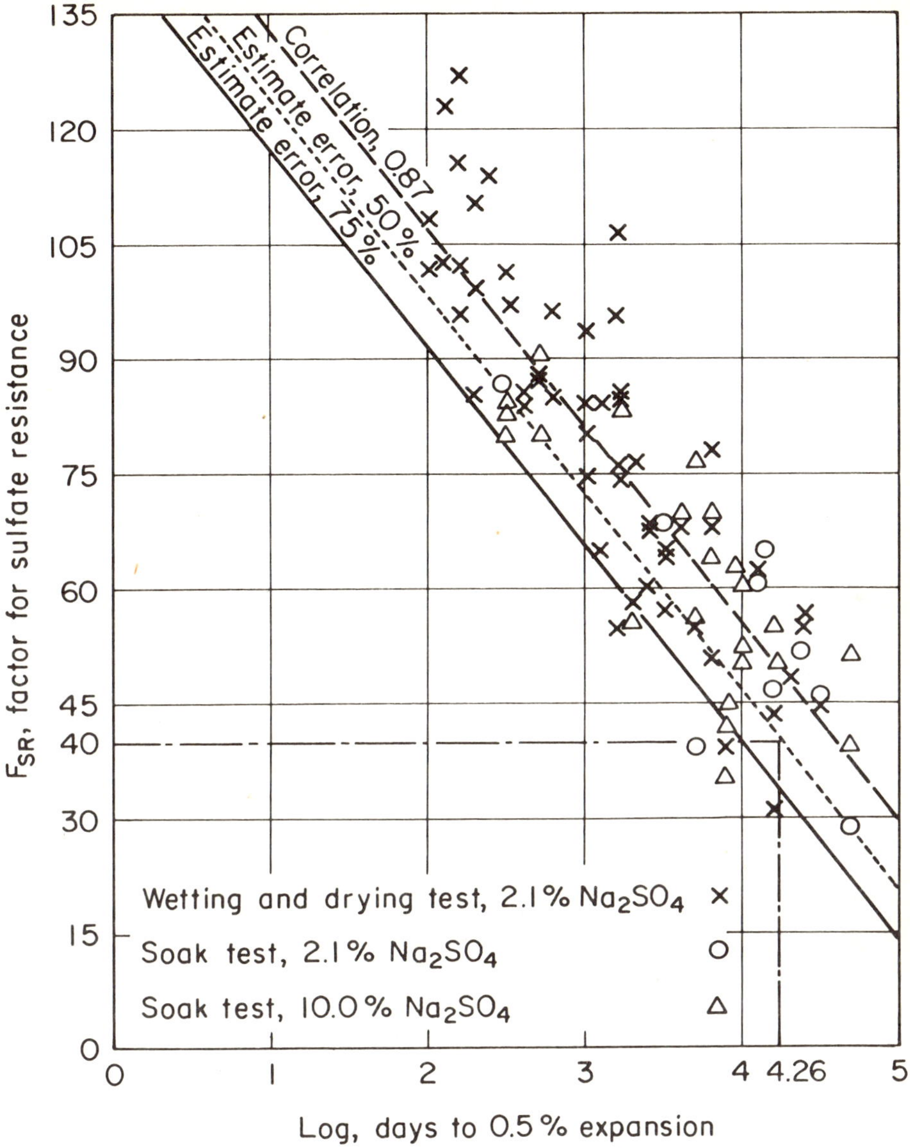

Fig. 11.1 Relationships between calculated sulfate resistance factors, F_{SR}, for various concretes and time to 0.5% expansion in tests of different severities.

Blended cements may not perform well in the Wolochow test, because at the time of testing the blended cement specimens will not usually be as well hydrated, strong, and impermeable as specimens made with portland cement. The accelerated test method adopted by ASTM in 1984 (ASTM C 1012) uses C 109 mortars cured one day at 35° C (95° F) and further cured at 23° C (74° F) to specified strength of 21 MPa (3000 psi), rather than specified curing times.[79b] Test data accumulated on this basis show improved performance for blended cements.[255]

Rosner et al.[257b] presented results of tests of mortars for sulfate resistance according to the procedures specified in C 1012 in mixed sulfate salt solutions containing 25.0 g of Na_2SO_4 and 43.4 g of $MgSO_4 \cdot 7H_2O$ per liter of solution. These mortars were made with one ASTM Type II portland cement with 1, 10, 15, and 20% replacements of portland cement by two Class F, one Class C, and one fly ash meeting Class F specifications except for available alkalies. The mortars were not proportioned exactly as specified in C 1012; the water to portland cement ratio increased from 0.46 to 0.51 as fly ash contents increased from 0 to 20%. The water to cement plus fly ash ratio decreased from 0.46 to 0.40, the water contents being systematically reduced and the sand contents being adjusted for flow with increasing fly ash contents. Weight, relative dynamic modulus, and porosity changes (by ethyl acetate vacuum saturation) were measured. In addition, length changes were determined at intervals from 7 to 360 days of soaking in sulfate solutions as specified in C 1012. The portland cement mortars tested as controls exceeded 0.1% expansion during 1 year in 3 out of 4 sets of data. The expansions of the mortars made with the three Class F fly ashes remained less than 0.05%; the Class C fly ash (21.3% CaO) mortars with 10% fly ash expanded even more after 150 days of soaking than the portland cement mortars; at 15 and 20% fly ash the expansions were slightly less than for the portland cement mortars.

Weight-change data showed that all specimens gained weight for 56 to 90 days and lost weight thereafter. The weight losses were least for the portland cement mortars (reaching a rate of about 5% per year after 180 days) and greatest (more than 8% per year) for mortars with the highest contents of the lowest CaO-content ashes. Relative dynamic modulus testing was discontinued after 150 days because surface scaling caused increasing testing difficulty.

Deterioration of the Type II portland cement mortars was to be expected because the sulfate concentrations used in this accelerated test are at least 23 times those for which ASTM Type II (moderately sulfate resistant) portland cements are recommended in the ACI Manual of Concrete Practice[24B] (150 to 1500 ppm SO_4 in water). They are at least 3.4 times as concentrated as those for which Type V (highly sulfate resistant) portland cements are recommended. Nevertheless, sulfate concentrations as high as these are still far below saturation with respect to the sulfate salts and are justified in testing because in field exposure conditions drying may increase the concentrations to those of saturated solutions. For such very severe exposure conditions (over 10,000 ppm SO_4 in water) Type V portland cement and pozzolans shown by test to improve sulfate resistance are recommended.

Several alternative test methods have been developed. Revay and Kovacs[258] have presented an accelerated test method using steam-cured, 10 × 40 × 160 mm specimens made with a gap-graded sand to ensure high porosity and permeability. Mehta and Gjorv[259] have developed a method in which the pH of the contact solution is automatically controlled by acid titration instead of being controlled by soluble constituents in the specimens. Deterioration caused by mildly acid sulfate attack is greatly accelerated by such procedures. The severity of attack, whether determined by loss of strength or by linear expansion of mortars made with a high C_3A content portland cement, increased with decreasing pH.[260]

Interpretation of the results of such accelerated tests of mortar bars in very aggressive solutions is not simple. Weight gains by hydration and reaction of calcium aluminates with sulfates from solution are to be expected. Weight losses by leaching and surface scaling may also occur. Expansion measurements indicate the expansion of the central core of the mortar bar which depends on both (1) expansive reactions near the surface causing tensile stress in the core and (2) expansive reactions in the core. The former may not affect the interior if the expansion is confined to a thin disintegrating layer at the surface. Measured expansions of 0.05%, if taken as elastic strains, indicate an average stress of perhaps 13 MPa (1900 psi); since the tensile strength of the mortar is much lower, the strains cannot be elastic but must indicate cracking.

Strength measurements in such tests increase initially because of hydration and may decrease both because of expansive disruption in the interior and reduction of size of specimens caused by surface deterioration. Relative modulus measurements need to be corrected for any changes of geometry or density to be accurate indicators of deterioration.

Results such as those reported by Rosner et al.[257b] suggest that the depth of penetration of sulfate solutions was more rapid into the portland cement control specimens, causing expansion in the core, but that the destruction of the paste matrix in the blended cement mortars caused more severe scaling of the mortar bar surfaces. The results of such performance tests depend strongly on how the permeabilities of the mortars are affected by different curing times. Allowance of additional curing times for some cements to reach a specified strength, as in ASTM C 1012, should improve the performance of those cements in the test. However, if builders do not control the time at which young concrete is exposed to aggressive environments, such test procedures may not be reliable indicators of the relative performance of different cements in practice.

It is apparent that testing for the relative sulfate resistance of different blended cements, or sulfate resistant portland cements, and predicting the performance of concretes from test results poses complex problems involving differences in relative rates of hardening at different temperatures as well as differences of composition and properties of hardened binders. This complexity is recognized in paragraph 3.4 of ASTM C 1012-84, which permits the user to modify the test method by use of other compositions of aggressive solutions, other ages or degrees of maturity as the basis for beginning the exposure, mortars of different proportions, or different or additional means of assessing the influence of such exposure. However, the need to set expansion limits will require specified test solutions for specification purposes.

Alkali-Aggregate Reaction Control

The history and status of ASTM standards relating to alkalies in cements, including tests for potential alkali reactivity of aggregates, have been critically reviewed.[64] Attention was focused on the need for research on a test method (such as ASTM C441) to rank cements in order of suitability for use with alkali-reactive aggregates.

The situation with respect to control of alkali-silica reactions in concrete with fly ash additions is somewhat similar to the case with sulfate resistance. A considerable body of empirical data which indicates the beneficial effects of some fly ashes already exists. However, questions have been raised concerning the validity of standard test procedures such as ASTM C227 and C441 that were designed to assure concrete performance in the field.[261,262] Whether fly ashes can be relied upon to counteract alkali-aggregate reactions in concrete remains a controversial issue.[263]

Powers and Steinour[264] examined the mechanism of the alkali-silica reaction and proposed a hypothesis concerning safe and unsafe reactions in such systems. They concluded that, in the presence of excess lime, a nonswelling lime-alkali-silica complex is formed as an equilibrium product (i.e., an alkali calcium silicate hydrate). However, under some conditions abnormal swelling occurs because an alkali-silica complex is formed which is lime deficient. The inability of lime to diffuse into the reactive particle where the alkali-silica gel is formed prevents formation of the nonswelling equilibrium product. Expansion is believed to result from swelling of an alkali-silica gel, most probably as a solid before cracking occurs, but it may also involve hydraulic pressure. The reactions proceed safely or with damage to the concrete according to the relative amounts of calcium and alkali ions in the outer part of the reacted layer of the siliceous particle. The swell pressures that can be developed in alkali-calcium-silica-water gels have been shown to be extremely sensitive to composition; experiments by Struble show distinct swell pressure peaks at specific compositions.[265]

If the quantity of reactive aggregate is very large or if the specific surface is very high, the available alkalies are distributed in low concentration (relative to calcium) over the reacting surfaces, and the expansion rate is low. If there is very little reactive aggregate or aggregate surface for reaction or alkali, the rate of expansion is also reduced. At a certain proportion, dependent on aggregate type and fineness, the expansion will be at a maximum; this is referred to as the "pessimum" proportion. Maximum expansion may occur with small proportions of highly reactive material. This is the reason for the 15% lower limit of pozzolan contents of ASTM Type P and IP cements and for the six tests for alkali reactivity of pozzolans in Type I(PM) cements in ASTM C595 (see Table 1.3).

Powers and Steinour proposed that the alkali in concrete may react with silica as long as any unreacted (reactive) silica remained because alkali was regenerated in a cyclic process. When the lime diffuses into the alkali-reacted zone of the silica particle and reacts with the alkali-silica complex, it releases some of the alkali to continue the attack on the silica. However, Kalousek's[(195)] work (see Fig. 7.4) and more recent work by Bhatty and Greening[(266)] have shown that the alkali retention capacity of the alkali calcium silicate hydrate at equilibrium is appreciable. The results in Fig. 11.2 also show that C-S-H with a C to S ratio of has a much higher alkali retention capacity than C-S-H with a C to S ratio of 1.25. The highest initial C to S ratio (1.32) C-S-H was apparently altered by the alkali solution during exposure. Initially it had the lowest alkali retention capacity, but this capacity rapidly increased from 28 to 90 days; at later ages it released alkali. Electron probe microanalysis of hydrated alite (C_3S) rims in portland cement pastes with and without fly ash have directly shown C to S ratios reduced from 1.71 to 1.55 in the pastes with fly ash.[(267)] This reduction was a result of an increase of the silicon content, not a decrease of the calcium content, even in the "inner product." A small increase of the potassium content was also found in the inner product; larger increases would be expected in the outer product where the C to S ratios are much lower (see Fig. 8.4).

The standard test method for determination of the effectiveness of mineral admixtures in preventing excessive expansion of concrete due to the alkali aggregate reaction (ASTM C441) is a mortar bar expansion test using a highly reactive (Pyrex glass) aggregate and a blend of the fly ash with high alkali cement.

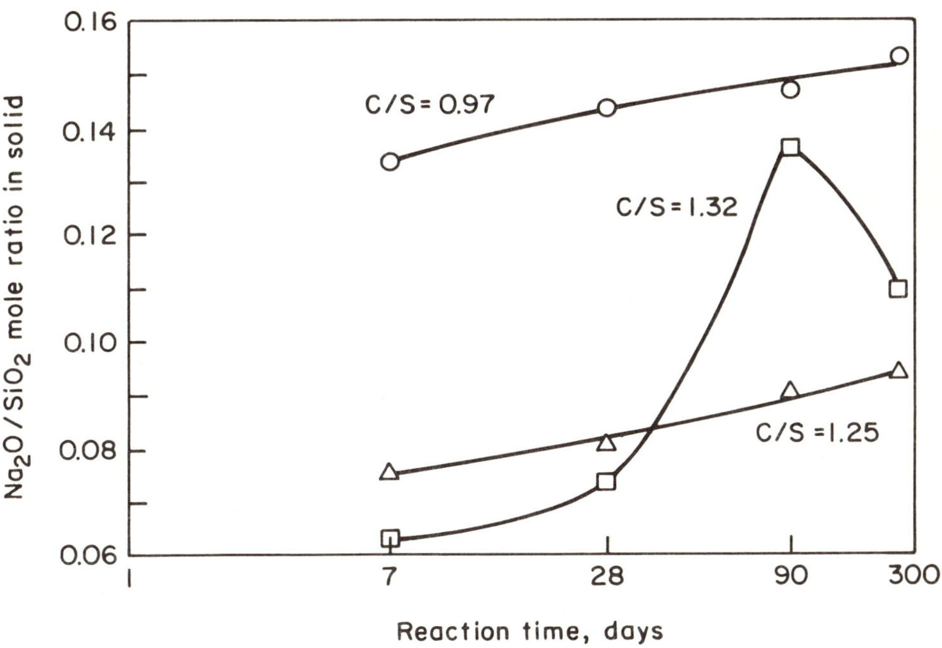

Fig. 11.2 Reaction of calcium silicate hydrates of different initial C/S ratios 1N NaOH solution. (Ref. 266)

Pepper and Mather[268,269] reported results of tests using similar methods in 1959. In tests of 20 pozzolans, including four different fly ashes, they showed that the minimum amount of fly ash required for effective control of the reaction was about 40% (34 to 52%) of the volume of the cement. Control was judged to be effective if a 75% reduction of the expansion was achieved. Berry and Malhotra[29] questioned whether the early strength losses caused by replacement of such large percentages of cement would be tolerable in most cases. However, it has been pointed out in the discussion of the paper[268] that the Pyrex glass aggregate used in these mortar tests tends to exaggerate the amount of pozzolan required relative to the amount required with most natural aggregates.

Pepper and Mather also correlated the amounts of each pozzolan required to control the expansions with their fineness and chemical characteristics. The mean particle diameter, the amount of silica dissolved in 1 N NaOH, and the alkali released in the presence of lime and water correlated well with the minimum percentage replacement of pozzolan for cement required for control. However, in a later report of work including results for high and low alkali content fly ashes, it was concluded that the regression equations were not valid, although the expansions did increase with alkali content of the fly ashes.[269] The between-batch and within-batch variances for many of the mixtures were found to exceed those specified in C441. It was concluded that partial drying due to leaky containers caused much of the variance but that other factors were probably involved as well.

The standard test method for assessing the potential alkali reactivity of cement-aggregate combinations (ASTM C227) has been criticized by Stark.[270] Data in Fig. 11.3 obtained with one of the most reactive aggregates encountered in the western United States (New Mexico aggregate containing reactive andesites and rhyolites) shows that only 20% fly ash was highly effective in control of these expansions. Even more importantly, Stark's data showed that even low alkali cements were not effective in preventing deleterious expansion in field concretes made with aggregates of this type. Although they would be judged innocuous by the failure criteria in ASTM C227 of less than 0.1% expansion after 6 months at 100° F (38° C) over water, numerous instances of serious map cracking of field concretes occurred in less than 15 years, bringing that failure criteria into question. It was further shown that the water-soluble alkali content in these concretes was much higher than could have been released from the low alkali cement used, indicating that most of the alkali came from the reacted aggregate itself. The control of this reaction by the use of fly ash was therefore all the more impressive.

Dunstan[271] has examined the effect of CaO contents in high-lime fly ashes on their ability to control alkali-aggregate reaction expansions. Small percentage additions of all fly ashes increased the expansions, but with higher percentages the expansions decreased. However, as the CaO content increased, the fly ash percentage required to reduce the expansion increased significantly.

Hobbs[262,272] has reviewed and extended earlier work in investigations of the influence of reactive aggregate content, mix proportions, cement alkali, and fly ash content on the time to cracking and expansion of mortar bars. Concerning

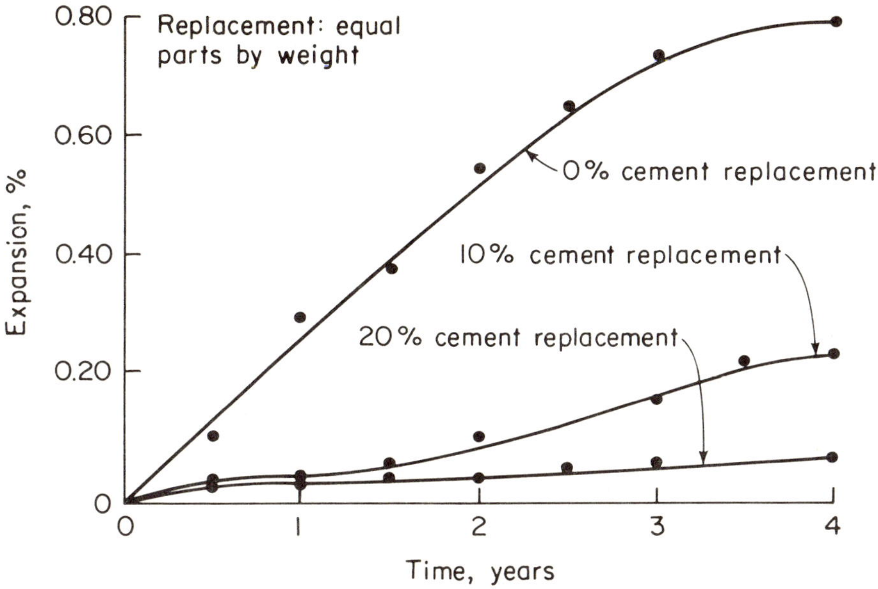

Fig. 11.3 Use of fly ash to control alkali-aggregate reaction expansions. ASTM C227 test results. Figure redrawn to include more recent PCA laboratory data. (Ref. 270)

the adequacy of ASTM C441 to measure the effectiveness of fly ashes in preventing expansion caused by alkali-silica reactions, Hobbs has raised questions concerning (1) the use of Pyrex glass as a reactive aggregate in mortar bar tests, (2) expansion limits that far exceed the cracking strains of mortars and concretes, (3) tests of cement-aggregate combinations (as in ASTM C227) at fixed proportions rather than those causing maximum expansions, and (4) determination of the amounts of fly ash actually required to prevent cracking and deleterious expansions.

For most of his experimental work Hobbs chose to use a reactive aggregate known as Beltane opal as being more representative than Pyrex glass of the porous opaline siliceous material present in many natural aggregates. Testing was done at 20° C with reactive aggregate contents adjusted to cement alkali contents to produce maximum expansions, rather than at 38° C with fixed cement-reactive aggregate ratios (as in ASTM C441). These "most critical" proportions were found to occur at a certain ratio of the water soluble Na_2O equivalent alkali content to the reactive silica content. For Beltane opal this is about 0.067. It was also found that there was a threshold ratio of 0.02, below which no cracking or expansion occurred. Hobbs found that the expansions correlated better with water-soluble alkali contents than with total alkali content. This is consistent with the finding that cracking and deleterious expansions may begin at very early ages (days). Hobbs' finding suggests that what happens early in the reaction predetermines the ultimate course of the reaction and that long-term tests should *not* be necessary to evaluate potentially deleterious combinations of materials.

Gutt et al.[263] determined the effect of fly ash on the reaction mixtures of C_3A, $Ca(OH)_2$, and K_2SO_4 to examine the mechanism whereby neutral alkali sulfates can increase the hydroxide content of a cement pore solution. The reaction without fly ash forms ettringite and/or calcium monosulfoaluminate, and the pore solution increases in alkalinity until it is almost completely converted to a KOH solution. When fly ash is included in the mixture, the same reactions occur but are greatly accelerated. These early reactions were suggested to be important in explaining the fact that pozzolans reduce the early expansions caused by reactive silica aggregate. On the other hand, for aggregates that release alkalies, the long-term effects could not be predicted from tests lasting 6 months or less.[261]

Only very small expansions (less than about 0.03% linear) occur before cracking. If the alkalies are supplied only by the portland cement and if the aggregate reacts rapidly, subsequent rapid expansions are a result of additional crack formation and growth, the rate decreasing with time as reactants are consumed. Cracking and expansions depend not only on the amount of gel produced by the reaction, but also on when it is produced, which in turn depends on the reactive aggregate content. At Beltane opal contents above about 20% of the total aggregate (in mortar), the concentration of alkalies in solution decreases rapidly to a threshold value during early stages of hardening, and gel formation produces negligible expansion. Hobbs developed a mathematical model which appears to account for most of the observed phenomena in this particular cement-reactive aggregate system in mortar bar tests.[273] He examined the use of seven different fly ashes (pfa's) in various mixtures with six different portland cements to control the cracking and expansion of mortar bars containing Beltane opal. Some results are shown in Fig. 11.4 for one portland cement (1.04% Na_2O equivalent alkalies) and seven blended cement (40% fly ash) mortars.

Expansions of the blended cement mortars increased rapidly after cracking, but the rate of expansion became negligible after a further 50 days. These long-term expansions varied by a factor of more than 7; even the least expansion (for pfa 7) was several times the cracking strain of about 0.03%. However, at a higher water to cement-plus-fly ash ratio (0.53), the expansion with only 30% of this same fly ash (pfa 7) was negligible, as was also the case with several other cement-fly ash combinations. Significantly, tests of pfa 7 with another portland cement of lower total alkali content showed cracking and expansion at both water to cement-plus-fly ash ratios.

Other results showed that cracking often occurred at earlier ages in the blended cement mortars than in the portland cement mortars, probably because of slower strength development of the blended cements and additional alkalies from the fly ashes. The fairly abrupt reduction in expansion rate observed at later ages appears to be a result of pozzolanic reactions of fly ashes which reduce the alkali concentration to the threshold level.

The results also indicated that the time to cracking and the long-term expansions (at 200 days) depend on the fineness and/or alkali contents of the fly ash.

The highest expansion occurred with the coarsest fly ash, and the lowest expansion occurred with the finest and lowest-alkali fly ash. Total alkali contents (as Na_2O) ranged from 1.30 to 3.67%; specific surface areas ranged from 235 to 460 m^2/kg. Hobbs found that no cracking or expansion occurred in portland cement mortars with total alkali contents below 3.5 kg/m^3. For mortars in which part of the cement was replaced by fly ash, the lowest alkali content (contributed by the portland cement) at which cracking occurred was 2.85 kg/m^3. If it was assumed that the fly ashes contribute about 0.2% alkalies by weight, the lowest alkali content at which cracking occurred was 3.4 kg/m^3, nearly the same as for the portland cement mortars. Hobbs concluded that fly ashes acted essentially as alkali diluters in reducing alkali-aggregate reaction expansions.

To investigate this conclusion directly, Gaze and Nixon[274] prepared mortar bars similar to those tested by Hobbs. They used a low-alkali portland cement with 10, 20, and 30% fly ash replacements and additions of potassium sulfate to maintain constant alkali concentrations. Tests were also done with mortars without fly ash but with alkali concentrations similarly adjusted to correspond to those that would have been obtained with fly ash replacements. Curing was done

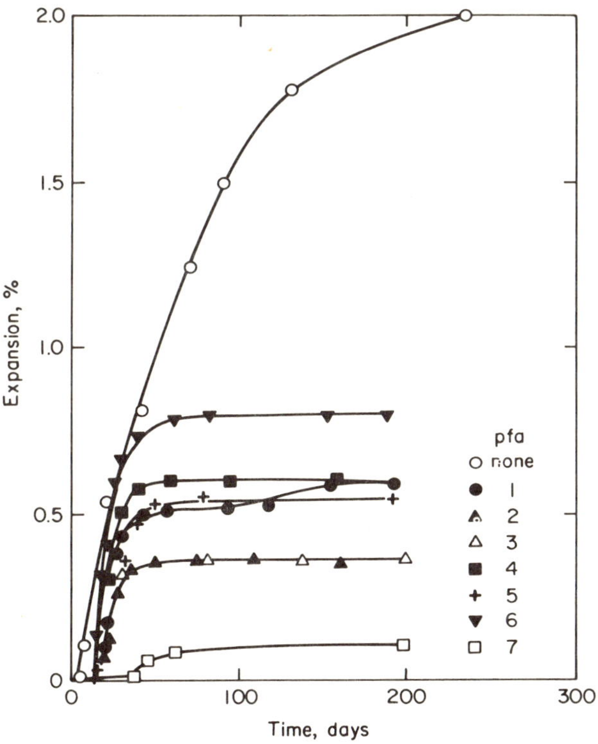

Fig. 11.4 Variation of expansion with age. w/(c + pfa) = 0.41, a/(c + pfa) = 2.75, pfa/(c + pfa) = 0.4. Blended cement content ~ 550 kg/m³.(Ref. 262)

at 38° C. For the mortars without fly ash, there was a considerable decrease in expansion with 20 or 30% reduction in alkali due to "dilution" alone. For the mortars with fly ash replacements, the decreases in expansion were even greater; the expansions were negligible at 20 and 30% fly ash even with the alkali content *not* reduced. It was concluded that the fly ash cannot be regarded only as an inert diluent, as might have been anticipated from the wide variations in the effects of different fly ashes shown in Fig. 11.4. However, Gaze and Nixon pointed out that these tests were carried out under conditions (38° C) that probably favored the effectiveness of the fly ash, which had good pozzolanic properties and a low-alkali content. They also stated that more information is required for practical use.

It is apparent that the rates of reactions and expansions caused by alkali-silica reactions in mortars and concretes depend in complicated ways on the properties of the reactive aggregates, cement, and fly ashes, as well as on test parameters such as mix proportions, particle size, humidity, and temperature. Research over many years confirms that some fly ashes can be very effective in reducing cracking and expansion when used with particular cement-aggregate combinations, but it also shows that the same fly ash may be less effective when used with another cement, reactive aggregate, or mix proportion. The main question concerns the validity of test methods used to evaluate fly ashes for this purpose.

The ASTM C 227 mortar bar test for cement-aggregate combinations has been criticized because reactions with aggregates that have been crushed to sand sizes are recognized to proceed differently than do those in concrete. The larger aggregate particles in concrete show characteristic outer reaction rims different from reacted material in their interiors. For some aggregates concrete prism or miniature rock prism tests for alkali reactivity are preferred over mortar bar tests.[63] However, such size factors may be of much less importance in tests that evaluate the effectiveness of fly ashes in preventing deleterious reactions if the most important reactions are those which reduce the alkali (and hydroxyl) ion concentrations to threshold levels. In that case, mortar bar tests, or even chemical tests, may be more rapid and efficient.

In the past it has been common practice to specify low-alkali cement for use with potentially reactive aggregates. Research and field experience have shown that this is not a sufficient remedy with certain aggregates and that it may be unnecessarily restrictive in many cases in which pozzolans are used. There is a clear need to improve our ability to predict the effectiveness that different fly ashes will have in controlling deleterious reactions.[275]

Such considerations suggest that there is still a need for research to improve understanding of these processes. A model for alkali-aggregate reaction control that works well enough to calculate the observed effects would permit prediction of the behavior from given parameters for any set of materials. A successful model or multivariant analysis would help to identify all of the parameters necessary to characterize fly ash, not only for this purpose, but also for the pozzolanic reaction contribution to strength, reduced permeability, frost resistance, and all porosity-dependent properties.

Frost Resistance

The frost resistance of concrete depends primarily on freezable water content and entrained air-void spacing factors. The freezable water content depends upon the degree of saturation and the capillary porosity, which in turn depend upon the water-cement ratio and degree of hydration, as discussed in "Porosity of the Hardened Cements" in Chapter 8. Hence, the rate of hydration and maturity of the concrete are important in reducing the freezable water content before testing or exposure to freezing temperatures. It is now well recognized that if fly ash concretes are not proportioned to achieve optimum water reduction and strength, the frost resistance at early ages may be reduced.[29] Characterization of fly ashes with respect to water requirements, reactivity, and strength development have been major topics in earlier chapters. Little has been done to relate fly ash characteristics to hardened cement pore size distributions, critical degree of saturation, salt scaling, and frost resistance.

Most research concerning the frost resistance of fly ash cement concretes now centers on the variable air-entraining agent requirements of fly ashes and the loss of entrained air during mixing and placement of the concrete. In "Air Entrainment" in Chapter 5 it was shown that the AE agent requirements of fly ashes could be determined from the Foam Index test and that loss of entrained air correlated well with the amount of easily decomposed organic matter in the fly ash.

Permeability and Durability

As discussed in "Effects of Fly Ash on Durability" in Chapter 1, the effects that fly ashes have on the permeability of concrete cause very large effects on its resistance to leaching, chemical attack, neutralization by carbonation, and penetration of chlorides to steel reinforcement.

Bakker[276] investigated the flow of water and ions through mortar and concretes with and without fly ash, using hollow cylinder specimens (containing salt solutions) immersed in a water bath. Flow was attributed to both osmotic and hydrostatic pressures. He concluded that all ordinary portland cement concretes are permeable to water but that during curing they become very impermeable to ions. He also concluded that blended cement concretes become impermeable to water as well as ions. However, in another study of the factors affecting chloride ion diffusion kinetics in hardened cements, blended cements were found to limit chloride ion diffusion to a far greater extent than plain portland cements.[277] Bakker's data may have been influenced by some other effects, including self-desiccation during hydration, which may not have been adequately considered in interpreting the results.

Feldman[221,222] made a systematic investigation of the effects of fly ashes on the porosity, pore structure, and durability of portland and blended cement pastes and mortars (see "Porosity of the Hardened Cements" in Chapter 8). He was able to show that some of the large pores (perhaps greater than about 0.09 μm) in blended cement pastes became blocked so completely during hydration that neither helium nor methanol (0.42 nm in diameter) could penetrate them. He also found in tests involving exposure to salt solutions (containing 27.5% $CaCl_2$, 3.9% $MgCl_2$, 1.8% NaCl, and 0.1% $NaHCO_3$) that the blended cements with fly ash were much more durable after 240 days of curing than two portland cements, although a Type I portland cement was more durable than the blended cement after only 14 days of curing. The improved durability of the well-cured blended cements was attributed to both the blocking of pores and the greatly reduced calcium hydroxide contents. However, it was not possible to demonstrate the actual differences in pore size distributions because of the damage done to the blended cement paste structure by mercury intrusion.

In connection with studies of corrosion of steel reinforcement, Kasai et al.[278] found that the air permeability and depth of carbonation of mortars usually increased with their fly ash contents if moist-cured for only short periods (7 days). Carbonation, oxygen, and chlorides may penetrate rapidly to the reinforcing steel if drying shrinkage cracks are large enough. However, when the cracks are of equal size, the rates of corrosion in concretes made with reactive pozzolan (silica fume) differ little from those in concretes without pozzolan.[279]

Protection of steel from corrosion does not appear to impose any additional chemical or physical requirements for fly ashes beyond those required for impermeability and continuity of the concrete cover in practice.

Mehta has proposed a simple test for evaluating the quality of fly ashes based on their ability to reduce the permeability of mortars to water during a 28-day curing period in sealed containers.[280] In tests of four fly ashes and four other pozzolans, he found considerable differences in the rates of water penetration into specimens during rewetting after they were dried in vacuum at 70° C. He attributed these differences to the effect of pozzolans on the volume of large pores. Three of the fly ashes reduced the penetration depth more than 20% when compared with controls made with inert admixtures. The coarsest fly ash (198 m^2/kg) yielded results no better than those for an inert admixture.

The differences in penetration depth obtained in this simple test were not nearly as great as might have been expected from the very large changes in permeability caused by hydration of cement in pastes[181,184] or fly ash replacements for portland cement in concretes as shown in Fig. 1.4. This finding suggests that the drying procedure greatly increased the permeabilities of all of the specimens and reduced such differences. If so, more moderate drying of the mortar may improve discrimination between different fly ashes.

Soundness, Durability, and Performance Prediction

The characterization of fly ashes for concrete soundness and durability is probably the most challenging of the technical problems, unless we choose to provide wide margins of safety. If materials are to be used in the most efficient ways, further work should be done to answer the questions raised concerning the autoclave expansion test and the tests for sulfate resistance and control of alkali-aggregate reactions. Further work should also be done to determine how to achieve the potential reduction of permeability of concrete by means other than unrealistically long curing periods for structural and paving concretes.

One of most significant characteristics of some fly ashes is their capacity to reduce the water requirement of concrete. This property makes it possible to achieve mix proportions with low porosities, low permeabilities, and very high strengths. However, low water-cement ratio mixtures are more sensitive to expansive reactions caused by free CaO, MgO, and sulfoaluminates than high water-cement ratio mixtures. If fly ashes are to be utilized most advantageously and without failures due to unsoundness, it will be necessary to characterize the fly ash so as to achieve a proper balance between the need for dense, impermeable concrete and the need to control expansive reactions. Proper characterization should limit the use of fly ashes that may cause unsoundness and will encourage the use of fly ashes which contribute most to soundness and durability.

To assure long-term durability of concretes that will be exposed to sulfate (and other aggressive) solutions in service we need to be able to predict from the results of performance tests or from materials characteristics, or both, that concretes made from those materials will have a satisfactory service life[(281)] under those conditions. However, even for portland cements, neither rapid test methods nor chemical compositions always rank different cements in the same order, nor do either always agree with experience. The inherent variability of fly ashes further complicates prediction of performance of concretes.

Prediction of concrete performance should entail more than pass/fail criteria for blended cements. What appears to be needed are predictive models incorporating input data reflecting both structure formation and deterioration processes in different formulations. Sufficiently accurate models can also be used to establish the materials characterization requirements and standard specifications for combinations of materials in concretes for service under different conditions.

Chapter 12
Summary

The following is a summary of the more important findings of this review of recent and older literature on fly ash and its use with portland cement. Critical evaluations of data and test procedures have been made and related to engineering properties and the underlying science of these materials. Some apparent conflicts, problems, and research needs have been identified. This chapter includes an assessment of priorities for future research.

Generation of Fly Ash

Study of the processes of formation of fly ashes provides much information on the composition and structure of fly ash particles and suggests what may be the most important causes of their variable effects when used in concrete. The major constituents in most fly ash particles are aluminosilica glasses formed during combustion from the melting of the inorganic constituents in the individual coal particles. During cooling, inorganic species in the vapor phase begin to condense as thin coatings (0.01 to 0.03 μm thick) on the larger particles and also as very fine primary particles (0.06 μm), which then grow by coagulation to 0.16 μm in the combustion zone for several seconds before being quenched. These particles and the surfaces of the larger particles are rich in inorganic components such as alkali sulfates and potentially toxic trace elements. Conditioning agents that improve the efficiency of electrostatic precipitators may introduce additional alkalies and other substances.

Organic residues of coal combustion, such as polycyclic aromatic hydrocarbons, are also adsorbed on the surfaces of the particles. These organic compounds may resist thermal and photochemical decomposition while adsorbed on the surfaces of the fly ash particles. This picture of the process suggests that, although the composition of the bulk material in the larger particles may vary widely, depending on the composition of the individual coal particles, the material deposited on the surfaces from the vapor phase may be much more uniform.

Surface characteristics of the fly ash particles are most important with respect to their effects on the properties of fresh concrete and its early strength development. Pozzolanic activity at later ages depends more upon the nature of the bulk material in the particles.

Fly Ash Characteristics That Affect Concrete Properties

Water Requirements of Fresh Mixtures

All fly ashes increase the water requirement for flow when added to cement paste or concrete, but some fly ashes reduce the water requirement when used as partial replacements for portland cement.

It is commonly believed that fly ash improves the workability of fresh concrete or reduces the water requirement for flow because of the spherical shapes of many of the particles. However, rheological studies of cement pastes with and without fly ash show that flocculation and dispersion effects are far more important. Some fly ashes have properties very similar to high-range water-reducing admixtures.

These studies suggest at least three important ways in which fly ash can influence the properties of fresh concrete. First, the organic compounds condensed on the fly ash particles during their cooling in the stack gases may function as water-reducing admixtures by dissolving and being readsorbed on both cement and fly ash particles.

Second, if the fly ash particle surfaces are all very similar, they will all have surface charges of the same sign. The silicate and aluminate phases in portland cements have surface charges of opposite signs (silicates are negative, aluminates positive). The fine particles are mostly monomineralic, so that they form very stable floc structures with particles of the opposite sign alternating in the floc structure. By contrast, particles of the same sign (such as calcium aluminate cements) are dispersed by mutual repulsion forces, and suspensions of such particles flow more easily. Fly ash with a (negative) surface charge will, in the course of mixing, come into contact with and adhere to the positively-charged aluminate and calcium-rich areas on the surfaces of the calcium silicates. The very fine fly ash particles may be regarded as being "adsorbed" by larger cement particles. If there are enough of these very fine particles, the cement particles will be effectively dispersed and will flow more easily. This hypothesis, a result of this review, has been independently confirmed by Chinese researchers.[(167)]

Third, alkali sulfates and other phases concentrated on the surfaces of fly ash particles are readily soluble and strongly influence the concentrations of ionic species in solution. There is some evidence that crystalline compounds grow from the vapor phase on the particles (or form by recrystallization later) in addition to being incorporated in the glass at the surface of the particles. Hence, the alkali and sulfate from the fly ash will enter into the cement solution equilibrium between calcium, alkalies, hydroxide, and sulfate ions. These concentrations then determine the course of the calcium aluminate and silicate hydration reactions.

Factors Affecting Pozzolanic Reactivity

Pozzolanic Reactions. There is not yet any general agreement on the exact mechanisms of the pozzolanic reaction and the alkali-silica reaction. Some research indicates that a surface-reacted silica layer separates from the fly ash particle to form an osmotic cell, which swells and is ruptured by osmotic pressure. The concentration of dissolved silica and the rate of reaction presumably depend upon the concentration of alkalies in solution. Strength development is delayed until the void under the surface-reacted layers is filled with solid reaction products.

The pozzolanic reactivity of fly ash particles with lime increases greatly with the concentration of alkalies in solution. Studies of the alkali-silica reaction have shown that the maximum swell pressures are reached in very narrow compositional ranges. There is as yet no quantitative model for these reactions.

The pozzolanic reaction between fly ash and the calcium hydroxide released by hydration of portland cement produces additional calcium silicate hydrates which contribute to strength development.

It has long been known that the C to S ratio of the calcium silicate hydrates produced by hydration of C_3S and C_2S averages about 1.5 under normal conditions. Recent research shows that calcium silicate hydrates have C to S ratios in this same range in the vicinity of cement particles but that in the vicinity of fly ash particles this ratio is significantly lower.

With respect to the rates of reaction, many studies have indicated that the pozzolanic reactions begin very slowly and that only after curing for several weeks does substantial reaction occur. Quantitative measurements of the degree of reaction of cement with fly ash show that alite hydration, after initial retardation, is accelerated by the presence of the fly ash, while aluminates are retarded. In addition, it has been found that the amount of calcium hydroxide produced by alite hydration is not reduced at early ages by the pozzolanic reaction. Instead, some low C to S ratio hydrates are produced on and near the fly ash particles. Because there is no reduction in the amount of calcium hydroxide, some researchers have erroneously concluded that there is no early reaction with the fly ash. It now appears that the early pozzolanic reaction reduces the average C to S ratio of the calcium silicate hydrate instead of reducing the amount of calcium hydroxide. At later ages nearly all of the calcium hydroxide may be consumed in the pozzolanic reactions.

The characteristics of fly ashes that most strongly influence pozzolanic reactivity are fineness, glass content, and silica or silica-plus-alumina content. Other factors are iron, calcium and carbon contents, and the composition of crystalline and organic substances on the particle surfaces. The variability of fly ashes, both in overall chemical composition and in the relative proportions of different particle types and fineness, results in extremely wide variations in pozzolanic reactivity and in cementing efficiency for strength development.

Most reviewers agree that it is not yet possible to predict the reactivity or pozzolanic activity (strength) from any known combination of physical and chemical measurements on the fly ash alone; thus, direct tests of strength development with lime or with portland cement are still necessary in order to evaluate fly ashes for use in blended cement or in concrete. In fact, even the pozzolanic activity tests, which are accelerated tests at elevated temperatures, are only very crude indicators of the performance of fly ashes at ambient temperatures and may not even rank different fly ashes in the proper order with respect to ordinary use.

Nevertheless, research to date has adequately demonstrated the importance of fineness, glass content, and silica-plus-alumina contents as major characterization variables. The interaction of these variables, together with other variables that can sometimes have large effects, makes the pozzolanic reaction behavior extremely complex. Linear regression analysis has not been successful in predicting behavior but may not have been very well explored. This situation exists partly because the true functional dependencies are probably nonlinear over the ranges of interest and also because much work to date has used characterization methods that are not satisfactory.

Fineness. Fineness has long been recognized as one of the most important characteristics of fly ashes, and the effects of variations of fineness have been extensively studied. Fineness has also proven to be one of the most difficult properties to measure satisfactorily. Several studies have shown that different methods can yield widely divergent results. Consequently, there is much confusion in the literature on the effects of fineness on the pozzolanic properties. However, certain procedures can be used to improve the utility of such measurements. For example, fly ashes have been washed to remove water-soluble substances, ignited to burn out carbon, magnetically separated to remove magnetite and hematite agglomerates, and wet-sieved to disperse agglomerates. One recent study has shown that even without such treatment Blaine specific surface measurements of fly ashes (from the same source) yielded better correlations with pozzolanic activity test results than the 45μm (No. 325) sieve residue, which is presently used as the fineness specification in ASTM C618 and C595. BET surface area measurements are also useful for indicating reactivity (Pozzolanic Index) and how much air-entraining agent or water-reducing admixture may be adsorbed by the unburned carbon and other porous particles. Methods of measuring the amounts and effects of the recently discovered submicron fraction need thorough investigation. There is still a need for a definitive study of how best to characterize the fineness and specific surface area of fly ash. Present ASTM specifications have been challenged in the literature.

Surface Characteristics. Water-soluble phases such as calcium sulfates and organics on the surfaces of fly ash particles can have important effects on early cement hydration reactions and on water requirements for flow. Fly ash may initially retard alite hydration, but it generally begins to accelerate alite hydration within the first 24 hours. Washing the fly ash eliminated the initial retardation. Other studies have shown C_3A hydration to be retarded by fly ash. Admixtures which accelerate the hydration of the silicates and retard the aluminate are highly desirable if the effects can be well controlled and predicted. These phenomena need to be better understood.

Recent research indicates that partial crystallization of glassy particles results in concentration of impurities on the particle surfaces, and this may also significantly influence the pozzolanic reactivity of the glass. Research of this nature requires expensive instrumentation, and such methods are probably not suitable for standard tests or specifications. Nevertheless, such phenomena must be recognized as a source of variability of fly ashes that may prove to be important.

Glass Content. Glass contents have been measured by dissolution in HF, by difference from QXRD of crystalline phases, and by optical microscopic counts. The most recent work indicates that reliable results can be obtained using dilute HF to completely dissolve the glass phases and that the rate of dissolution provides a good measure of pozzolanicity. Best results are obtained if other soluble phases are first removed with HCl. These techniques appear to be adequately developed so that they may be used without further extensive work.

Silica-Plus-Alumina Content. Silica-plus-alumina contents are also adequately measured by existing methods. The total SiO_2 + Al_2O_3 content appears to correlate very well with long-term pozzolanic activity. Although this finding seems well established, it is difficult to understand. It is generally believed that mullite and quartz, which form by partial crystallization of the glassy phases in the fly ash, are not reactive. The fact that the long-term pozzolanic activity correlates with the total SiO_2 + Al_2O_3 content, rather than just that in the glass, suggests that either the mullite and quartz do react or else they contribute to strength even if not reacted, e.g., as bonded aggregate particles. This matter does not appear to have been investigated and should be a part of future research.

Iron Content. Another closely related matter is the basis for inclusion of the Fe_2O_3 content in ASTM C618 chemical specifications for the total SiO_2 + Al_2O_3 + Fe_2O_3. It appears to be well established that in most fly ashes, most of the iron is present as unreactive hematite and magnetite. A small amount is present in the glass and seems to have a deleterious effect on its pozzolanic activity. Hence, it appears that iron should represent a negative component with respect to pozzolanic properties. If the specification was written merely as a matter of analytical convenience, then the matter should be reexamined in terms of current analytical practice.

Calcium Content. A somewhat similar question relates to the calcium oxide content of the fly ash. In low-calcium fly ashes (Class F), the calcium is mostly dissolved in the glass, except for some free CaO and surface deposits of calcium sulfates. The calcium in the glass appears to have a strong positive effect on reactivity, but this effect is complicated by the fact that calcium also promotes the formation of finer fly ash particles so that the fineness increases with calcium content. In high-calcium content fly ashes (Class C), the fly ash may also contain free CaO, C_3A, $CaSO_4$, $C_4A_3\overline{S}$, and other crystalline calcium compounds that have various effects.

Despite important differences between cementitious and pozzolanic fly ashes, there are no direct tests to distinguish between these properties of fly ashes specified in ASTM C618. CaO contents are not used in the specification, nor are there any differences in the pozzolanic activity test methods or specifications that distinguish cementitious from pozzolanic fly ashes. Hence, although Class C fly ashes are described in ASTM C618 as cementitious, there is nothing in the specification to ensure this, nor is there anything to prevent Class F fly ashes from being moderately cementitious. Since the calcium content has a large effect on the reaction chemistry, much of what has been learned about the effects of Class F fly ashes in reactions with cement must be reexamined with Class C fly ashes. Further research on sulfate resistance and alkali-aggregate reaction control will be required to determine how to characterize these fly ashes for such purposes.

Fly Ash-Cement Proportions and Reactivity. To compare results obtained by various investigators using different proportions of fly ash with lime or cement, the amounts required for complete reaction were estimated using typical pozzolanic fly ash oxide contents (50% SiO_2 and 30% Al_2O_3). These estimates showed that, for fly ash-calcium hydroxide mixtures, the proportions specified for the test for the Pozzolanic Activity Index with Lime (ASTM C 311) and the proportions used in many research investigations would be deficient in lime at advanced stages of reaction. Highly reactive pozzolanic fly ashes used in these proportions may produce low-lime hydrates different from those expected in use with portland cements in concrete and may yield lower strengths than less reactive fly ashes.

The amount of fly ash required to react completely with the calcium hydroxide produced by hydration of ordinary portland cements was similarly estimated. To obtain reaction products of the desired range of compositions at complete reaction requires 22 to 33% fly ash in mixtures with highly reactive pozzolanic fly ashes. ASTM C 311 for the Pozzolanic Activity Index with Portland Cement specifies about 30% fly ash and 70% cement by weight. This test should yield the desired reaction products and good strengths with highly reactive pozzolanic fly ashes. In concretes made with greater percentages of such fly ashes, loss of strength may occur at advanced stages of reaction because of partial decomposition of the calcium silicate hydrate.

Durability Considerations

One of the most important and urgent areas for future research concerns the characterization of fly ashes with respect to soundness and durability of the concrete. This area also presents some of the greatest opportunities to prevent problems with the use of fly ashes, especially those generated from lignite and subbituminous coals, and to actually improve the performance of concrete as a durable material.

Fly ash addition to cement has been shown to be effective in the control of unsoundness caused by excessive MgO as periclase in portland cement clinker. Cement plants would be able to use their limestone quarries more efficiently if this method of control could be used. Specification revisions would be required, which might prove very difficult because of the long time required for deleterious expansions of this kind to occur under ordinary conditions or to be demonstrated to be under control.

Potential unsoundness of blended-cement concretes resulting from the formation of ettringite from aluminates and sulfates also needs further investigation. The validity of accelerated cure paste and mortar expansion tests needs to be examined carefully for blended cements.

The use of some Class F fly ashes and curing to improve the sulfate resistance of well-cured concretes has been adequately demonstrated, but permeability and sulfate resistance are strongly dependent on curing time and conditions. Class C fly ashes, normally produced from lignite or subbituminous coal, are less effective or may even accelerate deterioration. Two approaches to assess the sulfate resistance of concretes containing fly ashes are being developed in the United States: (1) tests of cement blends by expansion measurements of mortar bars exposed to sulfate solutions after curing to specified strengths (not specified times) and (2) calculation of sulfate resistance factors from portland cement, fly ash, and blended cement compositions, water to cement ratios (but not curing time and conditions), and severity of exposure to which the concrete is to be exposed. The exact mechanisms, however, are not well enough understood to enable us to calculate and predict the progress of deterioration of concrete with certainty by either method. A validated model for the process is required to relate accelerated test data and materials characteristics to concrete performance. Such a model must include the effects of curing as well as the effects of fly ashes on the water to cement plus fly ash ratio required in particular concretes.

Control of deleterious alkali-silica reaction expansion with some fly ashes of low alkali and CaO contents has been successfully demonstrated. This controlling effect can be a very important benefit of fly ash in concretes. However, test methods are of limited validity in establishing the effectiveness of fly ashes of different compositions, and these methods need to be improved. Research on the mechanism of the expansion is necessary to predict performance with confidence.

Protection of steel reinforcement from corrosion by chloride salts or reduced pH resulting from carbonation by atmospheric carbon dioxide requires concretes of low permeability and continuity of concrete cover. Fly ashes can be used to produce dense concretes of low permeability when adequately cured, but surface drying shrinkage cracking must also be controlled.

Priorities for Future Research

At an EPRI workshop in 1982,[50B] cement and concrete technologists met with representatives of electric utilities and fly ash distributors to discuss R&D requirements to improve the use of fly ash in concrete. The underlying theme of the entire workshop was concern with the adequacy of test methods and ways of classifying fly ashes for successful use. The Panel on Short Term R&D Requirements[91] arrived at the following list of topics in order of descending priority:

1. General Methods of Evaluating Fly Ash for Suitability for Use in Concrete. This topic included work on many specific tests, including the ASTM Standard Test Methods discussed in Chapter 1 and elsewhere (see Rapid Tests for Fly Ash Quality and other specific items below). In addition, a Research System was proposed using reference samples (Standard Reference Materials) and coordinated interlaboratory testing and calibration such as is done by the NBS Cement and Concrete Reference Laboratory.

2. Methods of Proportioning Concrete to Compensate for the Effects of the Fly Ash.
   The panel recognized the need for additional work on the use of high CaO content (Class C) fly ashes. In addition, concern was expressed for proportioning not only for strength at a particular age at minimum initial cost, but also for long-term durability and other properties of the concrete. Hence, the proportioning R&D work was divided into two categories:

   a. For Strength, with emphasis on the outer limits of present usage and different types of materials.

b. For Durability, to ascertain that the concrete with fly ash is at least equivalent to portland cement concrete with regard to
Freezing and thawing
Alkali-aggregate reactions
Soft water attack
Seawater attack
Sulfate attack

3. Rapid Tests for Fly Ash Quality
The panel recognized that, if a set of very rapid and reliable tests were available at the power station to control the selection or diversion of fly ashes during operational changes, much variation in the quality of the fly ash selected for use with concrete might be avoided. Some of these tests already exist, but their significance with respect to performance is not yet understood. New methods are becoming available which may prove more useful. Some of these methods are

a. % Carbon—Leco Carbon Test (very rapid)
LOI—Differential Scanning Calorimeter

b. No. 325 screen residue—Alpine air jet sieve
Particle size distribution—L&N Microtrac and Micromeritics SediGraph
Specific surface area—Modified ASTM C204

c. Density

d. Color

e. Reactivity by conductivity in dilute HF (generally Class F only, as now used to obtain reactivity after 1 year in one hour)

f. Water soluble salts from ESP conditioners

4. Air Entrainment
Fly ash compositional variations are known to cause much variability in the air content or in the amount of air-entraining agents required to obtain good durability in freezing environments. Specific areas identified were

a. Limitations of the Foam Index Test

b. Effects of fly ashes on bubble size distributions

c. The development of guidelines for use of fly ash in air-entrained concrete

d. Alternative means of achieving air entrainment or frost durability in concrete with fly ash

e. Treatment of fly ash to prevent problems with air entrainment

5. Effect on Durability
In addition to the effects of fly ash on air entrainment for frost resistance of concrete, the proportioning of concrete to enhance durability, and the general methods of evaluating fly ash for suitability for use in concrete, the panel specifically indicated the need for additional information on test methods, fly ash characterization, and specifications concerning

a. Sulfate resistance

b. Alkali-aggregate reaction control

c. Soft water resistance

d. Sea water durability

6. Improved Tests for Activity
These include improved tests for pozzolanic reaction rates and strength development for the Class F fly ashes. There are also special tests to improve determination of the cementitious reactions and effects on properties of Class C fly ashes.

7. Effects of Fly Ash on Water Demand
This topic is actually a part of the first priority category on methods of evaluating fly ashes.

8. Effect of Chemical Admixtures
The panel also indicated that guidelines for the use of water-reducing and retarding chemical admixtures were required for use with fly ashes of different characteristics.

9. Curing Temperature
The standard tests lack flexibility in evaluating the performance of fly ash or concrete because of the use of fixed proportions and temperatures in pozzolanic activity index tests and at fixed room temperatures in concrete tests. Thermal activation at elevated temperatures, retardation at low temperatures, and the effects of alkalies also need to be evaluated.

Several authors cited in Chapter 1 have advocated revision of specifications for fly ashes so that chemical and physical characteristics are eliminated in favor of performance test results. However, we have seen that important performance tests presently in use are of questionable validity. Some are not based on a good understanding of the effects of fly ashes on the properties measured. In Chapters 5 and 8 we saw that standard test methods may yield misleading results. Water requirements may be much higher for some fly ashes in the ASTM mortar-flow test than in concrete where proportions may be adjusted to accommodate +325 mesh material. Highly pozzolanic fly ashes may not appear so in the ASTM pozzolanic activity index tests because of a high-water requirement or, in the test with lime, because the proportions yield lime-deficient hydrates. Optimization of cement-fly ash mixtures depends not only upon the factors which lead to maximum water reduction but also on test methods that determine proportions appropriate to the overall reaction chemistry. An even greater challenge for performance testing concerns tests for long-term soundness and durability of concrete, since these tests must yield results valid for the service life of concrete structures.

It should be relatively easy to improve short-term performance tests such as those for water requirement and pozzolanic activity, but performance tests for long-term durability on a strictly empirical basis are not only slow, but also uncertain. More basic research is necessary to define the mechanisms and parameters involved in strength development and deterioration processes and to establish limits for the parameters that will guarantee long-term performance.

The hydration and pozzolanic reactions directly affect structure formation, development of strength, and all other properties. Structure formation processes during portland cement hydration have long been described in terms of space-filling processes in the porous structure of cement pastes. Porosity measurements provide a quantitative means of correlating the extent of hydration with the development of strength and other properties. Uncertainties in the determination of the density and composition of the hydrated material and in the degree of reaction of the fly ash lead to uncertainties in calculation of the extent of space filling. Therefore, direct measurements of porosity are necessary, although methods used to date have not yet yielded very consistent results.

The basic gel-space ratio concepts discussed in "Structure and Porosity of Hardened Cements" in Chapter 6 and "Strength Development" in Chapter 8, however, should be applicable to the cement-fly ash system. Such an approach, together with our steadily improving conception of the paste microstructure, should make it possible to determine the true dependence of the rates of space filling and strength development on specific measurable properties of fly ashes. Ultimately, the correlation of such research with improved rapid tests should lead to the development of better tests to characterize fly ash, better performance of fly ash cements, less dependence on long-term tests to evaluate fly ashes, and more efficient use of resources.

References

1. ASTM C618-85, "Standard Specification for Fly Ash and Raw or Calcined Natural Pozzolan for Use as a Mineral Admixture in Portland Cement Concrete," pp. 385-388 in 22B.

2. Tuthill, L. H., "Mineral Admixtures," Chapter 46, pp. 804-822 in 15B.

3. Kokubu, M., "Fly Ash and Fly Ash Cement," pp. 75-105 in Part IV of 6B.

4. Davis, R. E.; Carlson, R. W.; Kelly, J. W.; and Davis, A. E., "Properties of Cements and Concretes Containing Fly Ash," *Proceedings, American Concrete Institute,* Vol. 33, pp. 577-612, 1937.

5. Faber, J. H., "U.S. Overview of Ash Production and Utilization," pp. 5-13 in 40B.

6. Helmuth, R. A.; Miller, F. M.; O'Connor, T. R.; and Greening, N. R., "Cement," in *Kirk-Othmer Encyclopedia of Chemical Technology,* 3rd ed., Vol. 5, pp. 163-193, New York, John Wiley & Sons, 1979.

7a. *Ash at Work,* Vol. XIII, No. 1, p. 3, National Ash Association, Washington, D.C., 1981.

7b. "Ash Production Ash Utilization—1984," American Coal Ash Association, Washington, D.C., August 1, 1985.

8. U.S. Environmental Protection Agency, "Guideline for Federal Procurement of Cement and Concrete Containing Fly Ash," *Federal Register,* Vol. 45, No. 226, pp. 76906-76921, November 20, 1980; Vol. 48, No. 20, pp. 4230-4253, January 28, 1983.

9. Bakker, W. T., "Use of Fly Ash in Cement and Concrete in the U.S.A.," pp. 3-80 to 3-89 in 50B.

10a. Frohnsdorff, G., and Clifton, J. R., "Fly Ashes in Cements and Concretes: Technical Needs and Opportunities," NBS IR81-2239, U.S. Dept. of Commerce, National Bureau of Standards, Washington, D.C., 27 pp.,1981.

10b. NRMCA Technical Information Letter No. 426, National Ready Mix Concrete Association, Silver Spring, Md., June 18, 1985.

11. Idorn, G. M., "Research and Development for the Use of Fly Ash in Cement and Concretes," pp. 2-1 to 2-33 in 50B.

12. Brown, P. W.; Berger, R. L.; Clifton, J. R.; and Frohnsdorff, G., "Limitations to Fly Ash Use in Blended Cements," pp. 518-529 in 40B.

13. Gutt, W., and Nixon, P. J., "Use of Waste Materials in the Construction Industry," *Materials and Structures,* No. 70, pp. 255-308, 1979.

14. Yeager, K., "Aqueous Discharge Monitoring and Treatment," *Electric Power Research Institute Journal,* Vol. 9, No. 4, pp. 44-49, 1984.

15. Minnick, L. J., "Development of Potential Uses for Residue from Fluidized Bed Combustion Processes," DOE-ERDA FE-2549-12-77, 42 pp., 1977.

16. McCreight, L. R.; Gatti, A.; and Rauch, H. W., "Feasibility Study of Coal Slag Based Glasses for Hot Gas Clean-Up," Final Report of Phase 1, General Electric Co., DOE-ERDA EF-77-C-01-2608, 57 pp., 1978 (NTIS TID 28660).

17a. Spellman, L. U., "Overriding Demand for Energy Conservation in the Cement Industry," *Ceramic Engineering & Science Proceedings,* Vol. 1, No. 9/10, pp. 963-967, 1980.

17b. Roy, Robert, "Energy Report U.S. Portland Cement Industry," 14 pp., Portland Cement Association, July 1986.

18. Portland Cement Association, pp. 20-26 in 10B.

19. Gordian Associates, "The Cement Industry," NATO/CCMS-46, 58 pp., Technical Information Center, Washington D.C., Energy Research and Development Admin., 1976.

20. Price, J. D.; Troop, P.; and Gershman, H. W., pp. 4-3, 4-10 to 4-15 in 44B.

21. National Materials Advisory Board, pp. 63, 92-94 in 18B.

22. Transcript, Public Hearing on Proposed Guideline for Federal Procurement of Cement and Concrete Containing Fly Ash, U.S. Environmental Protection Agency, Office of Solid Waste, SW-999, 56 pp., 1981.

23. Heffelfinger, J., "EPA's Position on Guidelines for Use of Fly Ash in Cement and Concrete," Presentation at Transportation Research Board, 59th Annual Meeting, Washington, D.C., Session 134 (taped), January 23, 1980.

24. Styron, R. W., "Quality Control and Beneficiation of Fly Ash," pp. 151-164 in 38B.

25. Leming, M., "Fly Ash in Ready Mixed Concrete, Management Considerations," pp. 14-1 to 14-13 in 53B.

26. Dunstan, E. R., Jr., "Fly Ash and Fly Ash Concrete," REC-ERC-82-1, Bureau of Reclamation, U.S. Department of the Interior, Denver, Colo., Engineering and Research Center, May 1984.

27. Dunstan, E. R., Jr., "Performance of Lignite and Subbituminous Fly Ash in Concrete—A Progress Report," Report No. REC-ERC-76-1, U.S. Bureau of Reclamation, 23 pp., 1976.

28. Kriner, R. W., Statement in Ref. 22, pp. 21-27, and 2nd attachment to Transcript.

29. Berry, E. E., and Malhotra, V. M., "Fly Ash for Use in Concrete—A Critical Review," *Proceedings, American Concrete Institute,* Vol. 77, No. 2, pp. 59-73, 1980. Discussion by Ashby (Ref. 154) and Abdun-Nur (Ref. 119).

30. Elfert, R. J., "Bureau of Reclamation Experiences with Fly Ash and Other Pozzolans in Concrete," pp. 80-93 in 39B.

31. Price, G. C., "Investigation of Concrete Materials for the South Saskatchewan River Dam," *Proceedings ASTM,* Vol. 61, pp. 1155-1179, 1961.

32. Lovewell, C. E., and Washa, G. A., "Proportioning Concrete Mixtures Using Fly Ash," *Proceedings, American Concrete Institute,* Vol. 54, No. 12, pp. 1093-1102, 1958.

33. Tynes, W. O., and Mather, B., "Investigation of Cement-Replacement Materials—Use of Large Amounts of Pozzolans in Lean Mass Concrete," Misc. Paper No. 6-123, Report 10, U.S. Army Engineers, Waterways Experiment Station, 27 pp., 1962. See also Bryant, Mather, "Use of Concrete of Low Portland Cement Content in Combination with Pozzolans and Other Admixtures in Construction of Concrete Dams," *Proceedings, American Concrete Institute,* Vol. 71, No. 12, pp. 589-599, 1974.

34. Smith, I. A., "The Design of Fly Ash Concretes," *Proceedings, Institution* of Civil Engineers, London, Vol. 36, pp. 769-790, 1967.

35. Cannon, R. W., "Proportioning Fly Ash Concrete Mixes for Strength and Economy," *Proceedings, American Concrete Institute,* Vol. 65, No. 11, pp. 969-979, 1968.

36. Cannon, J. P., and Murti, G. R. K., "Concrete Optimized Mix Proportioning (COMP)," *Cement and Concrete Research,* Vol. 1, pp. 353-366, 1971.

37. Dodson, V. H., "The Strength Contribution of Fly Ash Concrete—A New Approach to Its Estimation," pp. 184-193 in 49B.

38. Lovewell, C. E. and Hyland, E. J., "A Method of Proportioning Structural Concrete Mixtures with Fly Ash and Other Pozzolans," *Proportioning Concrete Mixes,* ACI SP-46, pp. 108-140, Detroit, American Concrete Institute, 1974.

39. Rosner, J. C., "Let's Design Fly Ash Concretes, Not Compare Them," pp. 560-572 in 40B.

40. Ghosh, R. S., "Proportioning Concrete Mixes Incorporating Fly Ash," *Canadian Journal of Civil Engineering, Ottawa,* Vol. 3, pp. 68-82, 1976.

41. Owens, P. L., "Fly Ash and Its Usage in Concrete," *Concrete,* Vol. 13, No. 7, pp. 21-26, 1979.

42. Styron, R. W., "Quality Control of Fly Ash—The Combustion Process," *Proceedings of the Seventh Kentucky Coal By-Products Seminar,* Lexington, Kentucky, pp. 12-21, June 1981.

43a. Munday, J. G. L.; Ong, L. T.; and Dhir, R. K., "Mix Proportioning of Concrete with PFA: A Critical Review," pp. 267-288 in Vol. 3, 54B.

43b. Majko, R. M., and Pistilli, M. F. "Optimizing the Amount of Class C Fly Ash in Concrete Mixtures," *Cement, Concrete, and Aggregates,* CCAGDP, Vol. 6, No. 2, pp. 105-119, 1984.

44. ACI Committee 211, "Standard Practices for Selecting Proportions for Normal, Heavyweight, and Mass Concrete," ACI 211.1-81, pp. 211.1-1 to 32 in 24B.

45. Dodson, V. H., and Roberts, L. R., "Portland Cement-Fly Ash-Aggregate Concretes," U.S. Patent No. 4,210,457, July 1, 1980.

46. Lane, W. O., "Effects of Fly Ash on Freshly Mixed Concrete," *Concrete International,* Vol. 5, No. 10, pp. 50-52, 1983.

47. Dodson, V. H., "The Effect of Fly Ash on the Setting Time of Concrete—Chemical or Physical," pp. 166-171 in 49B.

48. Philleo, R. E., "Fly Ash in Mass Concrete," pp. 69-79 in 39B.

49. Owens, P. L., "The Reactions of Fly Ash and Portland Cement with Relation to the Strength of Concrete as a Function of Time and Temperature," Vol. III, pp. IV-60-65, in 17B.

50. Tynes, W. O., "Investigation of Water-Reducing Admixtures for Concrete," Technical Report No. 6-584, U.S. Army Engineers, Waterways Experiment Station, 20 pp., 1961.

51. Samarin, A., and Ryan, W. G. J., "Experience in Use of Admixtures in Concrete Containing Cement and Fly Ash," Workshop on the Use of Chemical Admixtures in Concrete, University of New South Wales, Kensington, NSW, Australia, pp. 91-112, 1975.

52. Lane, R.O., and Best, J.F., "Laboratory Studies on the Effects of Superplasticizers on the Engineering Properties of Plain and Fly Ash Concretes," pp. 379-402 in 16B.

53. Mukherjee, P. K.; Loughborough, M. T.; and Malhotra, V. M., "Development of High-Strength Concrete Incorporating Large Percentage of Fly Ash and Superplasticizers," *Cement, Concrete, and Aggregates* ASTM, Vol. 4, No. 2, pp. 81-86, 1983.

54. Larson, T. D., "Air Entrainment and Durability Aspects in Fly Ash Concrete," *Proceedings, ASTM,* Vol. 64, pp. 866-886, 1964.

55. Klieger, P., and Perenchio, W. F., "Laboratory Studies of Blended Cements: Portland Pozzolan Cements," Portland Cement Association R&D Bulletin RD013, 10 pp., 1972.

56. Perenchio, W. F., and Klieger, P., "Further Laboratory Studies of Portland-Pozzolan Cements," Portland Cement Association R&D Bulletin RD041, 9 pp., 1976.

57. Gebler, S. H., and Klieger, P., "Effect of Fly Ash on the Air-Void Stability of Concrete," pp. 103-119 in 54B.

58. Lamond, J. F., "Twenty-Five Years Experience Using Fly Ash in Concrete," pp. 47-69 in 54B.

59. Hansen, W. C., "The Chemistry of Sulfate-Resisting Portland Cements," pp. 18-55 in 4B.

60. Mather, B., "Field and Laboratory Studies of the Sulfate Resistance of Concrete," pp. 66-76 in 4B.

61. Kalousek, G. L.; Porter, L. C.; and Benton, E. J., "Concrete for Long-Time Service in Sulfate Environment," *Cement and Concrete Research,* Vol. 2, No. 1, pp. 79-89, 1972.

62a. Dunstan, E. R., Jr., "A Possible Method for Identifying Fly Ashes That Will Improve the Sulfate Resistance of Concretes," *Cement, Concrete, and Aggregates,* ASTM, Vol. 2, No. 1, pp. 20-30, 1980.

62b. Dunstan, E. R., Jr., "Fly Ash and Fly Ash Concrete," REC-ERC-82-1, Bureau of Reclamation, U.S. Department of the Interior, Denver, Colo., Engineering and Research Center, May 1984.

63. Grattan-Bellew, P. E., "Study of Expansivity of a Suite of Quartzwackes, Argillites, and Quartz Arenites," pp. 113-140 in 12B.

64. Frohnsdorff, G.; Clifton, J. R.; and Brown, P. W., "History and Status of Standards Relating to Alkalies in Hydraulic Cements," pp. 16-34 in 14B.

65. Swenson, E. G., and Gillott, J. E., "Characteristics of Kingston Carbonate Rock Reaction," Highway Research Board Bulletin 275, pp. 18-31, 1960.

66. RILEM 12-CRC Committee, "Corrosion of Reinforcement in Concrete," *Materials and Structures,* Vol. 9, No. 51, pp. 187-206, 1976.

67. Rasheeduzzafar, Dakhil, F. H., and Al-Gahtani, A. S., "Deterioration of Concrete Structures in the Environment of the Middle East," *Proceedings, American Concrete Institute,* Vol. 81, No. 1, pp. 13-20, 1984.

68. Larsen, T. J., and Page, G. C., "Fly Ash for Structural Concrete in Aggressive Environments," pp. 573-587 in 40B.

69. Tsukayama, R.; Abe, H.; and Nagataki, S., "Long-Term Experiments on the Neutralization of Concrete Mixed with Fly Ash and the Corrosion of Reinforcement, Vol. III, pp. IV-30 to 35 in 17B.

70. Ho, D. W. S., and Lewis, R. K., "Carbonation of Concrete Incorporating Fly Ash as a Chemical Admixture," pp. 333-346 in 54B.

71. Lewandowski, R., "Effect of Different Fly Ash Qualities on the Properties of Concrete, Part 3," *Betonwerk und Fertigteil-Technik,* Heft, March 1983, pp. 152-158, 1983 (in German).

72. Manz, O., "Lignite Production and Utilization," pp. 39-57 in 40B.

73. ASTM C311-85, "Standard Methods of Sampling and Testing Fly Ash or Natural Pozzolans for Use as a Mineral Admixture in Portland Cement Concrete," pp. 236-243 in 22B.

74. ASTM C595-86, "Standard Specification for Blended Hydraulic Cements," pp. 373-381 in 21B.

75. Manz, O., "American and Foreign Characterization of Fly Ash for Use in Concrete," pp. 269-279 in 49B; see also pp. 235-245 in 51B.

76. Smith, M. A., "Seven National Standards for Pulverized Fuel Ash Compared," *Precast Concrete,* Vol. 8, No. 4, pp. 191-193; No. 5, pp. 257-258, 1977.

77. Manz, O. E., "Review of International Specifications for Use of Fly Ash in Portland Cement Concrete," pp. 187-200 in 54B.

78. Rossouw, E., and Krüger, J., "Review of Specifications for Additives for Use in Concrete," pp. 201-220 in 54B.

79a. Mather, K., "Tests and Evaluation of Portland and Blended Cements for Resistance to Sulfate Attack," pp.74-86 in 14B.

79b. ASTM C1012-84, "Standard Test Method for Length Change of Hydraulic-Cement Mortars Exposed to a Mixed Sodium and Magnesium Sulfate Solution," pp. 607-611 in 21B.

80. Butler, W. B., "A Critical Look at ASTM C618 and C311," *Cement, Concrete, and Aggregates,* ASTM Vol. 4, No. 2, pp. 68-72, 1982; see also pp. 199-213 in 51B.

81. McKerall, W. C., and Ledbetter, W. B., "Variability and Control of Class C Fly Ash," *Cement, Concrete, and Aggregates,* ASTM, Vol. 4, No. 2, pp. 87-93, 1982.

82. Mielenz, R. E., "Specifications on Fly Ash for Use in Concrete: Where Are We—Where Should We Go?", pp. 482-495 in 46B.

83. Halstead, W. J., "Quality Control of Concrete Containing Fly Ash," Report VHTRC 81-R38, Charlottesville, Virginia, Virginia Highway and Transportation Research Council, 1981, reprinted by National Ready Mix Association as NRMCA Publication No. 164, 1981.

84. Cain, C. J., "Utilization of Fly Ash in Cement and Concrete," pp. 4-14 to 4-25 in 50B.

85. Miller, W., "The Evolution of Fly Ash Specifications," pp. 4-01 to 4-06 in 53B.

86. Werner, O., "Current Specifications Requirements and Needed Changes," pp. 5-01 to 5-10 in 53B.

87. Kocuvan, I., "A Proposal for Classification of Fly Ashes," pp. 215-224 in 46B.

88. Nicolescu, L. D., "Criterion of Estimating the Efficiency of Fly Ashes in Cements and Concretes," pp. 624-630 in 46B.

89. Cain, C. J., "Effects of Various Types of Fly Ash on Behavior and Properties of Concrete," pp. 260-268 in 49B.

90. Diamond, S., "Long-Term R&D Requirements," pp. 4-34 to 4-44 and pp. 6-14 to 6-17 in 50B.

91. Helmuth, R., "Short-Term R&D Requirements," pp. 4-26 to 4-33 and pp. 6-7 to 6-13 in 50B.

92. Vorres, K., "Coal" in *Kirk-Othmer Encyclopedia of Chemical Technology,* 3rd ed., Vol. 6, pp. 224-282, New York, John Wiley & Sons, 1979.

93. ASTM D388-84, "Standard Classification of Coals by Rank," *1985 Annual Book of ASTM Standards,* Vol. 05.05, pp. 247-252, Philadelphia, ASTM, 1985.

94. ASTM D3172-73, "Standard Method for Proximate Analysis of Coal and Coke," *1985 Annual Book of ASTM Standards,* Vol. 05.05, pp. 403-404, Philadelphia, ASTM, 1985.

95. Spackman, W.; Davis, A.; Walker, P. L.; Lovell, H. L.; Essenhigh, R. H.; Vastola, F. J.; and Given, P. H., "Characteristics of American Coals in Relation to Their Conversion into Clean Energy Fuels," Quarterly Progress Report Fe-2030-3, Energy Research and Development Administration, Pennsylvania State University, 1976.

96. Murtha, M. J., "The Mining and Utilization of Western Coal," Report to U.S. DOE by Ames Laboratory, IS-4616, Iowa State University, 1979.

97. Smoot, L. D., "General Characteristics of Coal," pp. 123-132 in 30B.

98. Perry, R. H.; Chilton, C. H.; and Kirkpatrick, S. D., "Combustion of Coal," *Chemical Engineers' Handbook,* New York, McGraw-Hill, 1963.

99. Flagan, R. C., "Ash Particle Formation in Pulverized Coal Combustion," Paper No. 77-4, Spring Meeting of the Western States Section, Seattle, Washington, The Combustion Institute, 1977.

100. Malte, P. C., and Rees, D. P., "Mechanisms and Kinetics of Pollutant Formation During Reaction of Pulverized Coal," Chapter 11 in 30B.

101. Padia, A. S.; Sarofim, A. F.; and Howard, J. B., "The Behavior of Ash in Pulverized Coal Under Simulated Combustion Conditions," paper presented at the Spring Meeting of the Central States Section, Columbus, Ohio, The Combustion Institute, 1976.

102. Field, M. A.; Gill, D. W.; Morgan, B. B.; and Hawksley, P. G. W., in 28B.

103. Raask, E., "Cenospheres in Pulverized-Fuel Ash," Institute of Fuel, Vol. 41, pp. 339-344, 1968.

104. Ulrich, G. D., "Theory of Particle Formation and Growth in Oxide Synthesis Flames," *Combustion Science Technology,* Vol. 4, pp. 47-57, 1971.

105. Ulrich, G. D.; Milnes, B. A.; and Subramanian, N. S., "Particle Growth in Flames, II. Experimental Results for Silica Particles," *Combustion Science Technology,* Vol. 14, pp. 243-249, 1976.

106. Ulrich, G. D., "Investigation of the Mechanism of Fly Ash Formation in Coal-Fired Utility Boilers," Quarterly Report to DOE EX-76-C-01-2205, 32 pp., Durham, New Hampshire University, June 1978.

107. Linton, R. W.; Williams, P.; and Evans, C. A., Jr., "Determination of the Surface Predominance of Toxic Elements in Airborne Particles by Ion Microprobe Mass Spectrometry and Auger Electron Spectrometry," *Analytical Chemistry,* Vol. 49, No. 11, pp. 1514-1521, 1977.

108. Natusch, D. F. S., "Characterization of Fly Ash from Coal Combustion," Report No. CONF-780481-3, Workshop on Measurement Technology and Characterization of Primary Sulfur Oxides Emission from Combustion Sources, Southern Pines, N.C., 1978.

109. Hansen, L. D., and Fisher, G. L., "Elemental Distribution in Coal Fly Ash Particles," *Environmental Science and Technology,* Vol. 14, No. 9, pp. 1111-1117, 1980.

110. Korfmacher, W. A.; Wehry, E. L.; Mamantov, G.; and Natusch, D. F. S., "Resistance to Photochemical Decomposition of Polycyclic Aromatic Hydrocarbon Vapor Adsorbed on Coal Fly Ash," *Environmental Science and Technology,* Vol. 14, No. 9, pp. 1094-1099, 1980.

111. Gallaer, C. A., "Flue-Gas Conditioning," Section 7.1 in *Electrostatic Precipitator Reference Manual,* EPRI:CS 2809, Palo Alto, California, Electric Power Research Institute, 1983.

112. Meininger, R. C., NRMCA Technical Information Letter No. 381, Silver Spring, Maryland, National Ready Mixed Concrete Association, 1981.

113. Castle, G. S. P., "Mechanisms Involved in Fly Ash Precipitation in the Presence of Conditioning Agents," IEEE Transactions on Industry Applications, Vol. 14-16, No. 2, pp. 297-301, 1980.

114. Ravina, D., "Production and Collection of Fly Ash for Use in Concrete," pp. 2-11 in 49B.

115. Power, J. P., "Production of Fly Ash," pp. 4-1 to 4-13 in 50B.

116. Hyland, E. J., "Fly Ash Quality Control: Important to Users," *Modern Concrete,* Vol. 43, No. 6, pp. 60-62, 1979.

117. Styron, R. W., "Benefication and Quality Control of Fly Ash," 10 pp. in 47B; see also Ref. 24.

118. Mouritzen, J., "Fly Ash in the Cement and Concrete Industry," *Nordisk Betong,* Vol. 2, pp. 27-36, 1980.

119. Abdun-Nur, E. A., Discussion of Ref. 29, *Proceedings, American Concrete Institute,* Vol. 78, No. 1, p. 86, 1981.

120. Nielsen, H. C. A., "Preparation of Fly Ash Cements," Vol. III, pp. IV-72-77 in 17B.

121. Luke, W. I., "Nature and Distribution of Particles of Various Sizes in Fly Ash," U.S. Army Engineers, Waterways Experiment Station, Technical Report No. 6-583, 21 pp., 1961.

122. Minnick, L. J., "Fundamental Characteristics of Pulverized Coal Fly Ashes," *Proceedings, ASTM,* Vol. 59, pp. 1155-1177, 1959.

123. Watt, J. D., and Thorne, D. J., "Composition and Pozzolanic Properties of Pulverized Fuel Ashes, I. Composition of Fly Ashes from Some British Power Stations and Properties of Their Component Particles," *Journal of Applied Chemistry,* Vol. 15, pp. 585-594, 1965; see also Refs. 219, 220.

124. Diamond, S., "The Characterization of Fly Ashes," pp. 12-23 in 49B.

125. Diamond, S., and Lopez-Flores, F., "On the Distinction in Physical and Chemical Characteristics Between Lignitic and Bituminous Fly Ashes," pp. 34-45 in 49B.

126. Adams, L. D., "Problems Involving the Analysis and Use of Fly Ash in the Cement Industry," Paper 9-T-80P, presented at American Ceramic Society Pacific Coast Regional Meeting, October 1980.

127. Sersale, R., "Structure and Characterization of Pozzolanas and of Fly Ashes," Vol. I, pp. IV-1/3-18 in 17B.

128. Adams, L. D., "X-Ray Diffraction Analysis of Portland Cement and Related Materials—A Quality Control Tool at Southwestern Portland Cement Co. for 12 Years," pp. 261-264, *Proceedings of the Fifth International Conference on Cement Microscopy,* March 1983, Nashville, Tennessee, International Cement Microscopy Association, 278 pp., 1983.

129. Scheetz, B. E.; Strickler, D. W.; Grutzeck, M. W.; and Roy, D. M., "Physical and Chemical Behavior of Selectively Etched Fly Ashes," pp. 24-33 in 49B.

130. Diamond, S., "On the Glass Present in Low-Calcium and High-Calcium Fly Ashes," *Cement and Concrete Research,* Vol. 13, pp. 459-464, 1983.

131. Fisher, G. L.; Prentice, B. A.; Silberman, D.; Ondov, J. M.; Bierman, A. H.; Ragaini, R. C.; and McFarland, A. R., "Physical and Morphological Studies of Size-Classified Coal Fly Ash," *Environmental Science and Technology,* Vol. 12, No. 4, pp. 447-451, 1978.

132. Fisher, G. L.; Chang, D. P. Y.; and Brummer, M., "Fly Ash Collected from Electrostatic Precipitators: Microcrystalline Structures and the Mystery of the Spheres," *Science,* Vol. 192, No. 4239, pp. 553-555, 1976.

133. Carpenter, R. L.; Clark, R. D.; and Su, Y. F., "Fly Ash from Electrostatic Precipitators: Characterization of Large Spheres," *Journal of the Air Pollution Control Association,* Vol. 30, No. 6, pp. 679-681, 1980.

134. Powell, R. A., and Spicer, W. E., pp. 85-109 in 34B.

135. Hulett, L. D., and Weinberger, A. J., "Some Etching Studies of the Microstructure and Composition of Large Aluminosilicate Particles in Fly Ash from Coal-Burning Power Plants," *Environmental Science and Technology,* Vol. 14, No. 8, pp. 965-970, 1980.

136. Hulett, L. D.; Weinberger, A. J.; Northcutt, K. J., and Ferguson, M., "Chemical Species in Fly Ash from Coal-Burning Power Plants," *Science,* Vol. 210, pp. 1356-1358, Dec. 19, 1980.

137. Entin, Z. B.; Lepeshenkova, G. G.; and Yashina, E. T., "Hydration and Hardening of Portland Cements with Ash," *Tsement,* Vol. 1, pp. 15-16, 1976 (in Russian); see also Ref. 127.

138. Sharkey, A. G.; Schultz, J. L.; White, C.; and Lett, R., "Analysis of Polycyclic Organic Material in Coal, Coal Ash, Fly Ash, and Other Fuel Emission Samples," Final Report, EPA/600/2-76/075, 31 pp., Pittsburgh Energy Research Center, 1976.

139. Natusch, D. F. S., "Formation and Transformation of Polycyclic Organic Matter from Coal Combustion," Report to DOE, Contract EE-77-S-02-4347, 27 pp., Fort Collins, Colorado State University, 1978.

140. Rixom, M. R., pp. 11-12 in 13B.

141. Minnick, L. J.; Webster, W. C.; and Purdy, E. J., "Predictions of the Effect of Fly Ash in Portland Cement Mortar and Concrete," *Journal of Materials,* ASTM, Vol. 6, No. 1, pp. 163-187, 1971.

142. Hobbs, D. W., "The Effect of Pulverized-Fuel Ash Upon the Workability of Cement Paste and Concrete," *Magazine of Concrete Research,* Vol. 32, No. 113, pp. 219-226, 1980. Discussion by Helmuth (Ref. 164).

143. Brink, R. H., and Halstead, W. J., "Studies Relating to the Testing of Fly Ash for Use in Concrete," *Proceedings, ASTM,* Vol. 56, pp. 1166-1206, 1956.

144. Chopra, S. K., and Narain, S. N., "Measurement of the Specific Surface of Fly Ash," *Materials Research and Standards,* ASTM, Vol. 4, No. 9, pp. 487-490, 1964.

145. Cabrera, J. G., and Gray, M. N., "Specific Surface, Pozzolanic Activity and Composition of Pulverized Fuel Ash," *Fuel,* Vol. 52, pp. 213-219, 1973.

146. Alexander, K. M., "Observations on the Blaine Method for Determining Fineness and on the Relationship Between Surface Area and Pozzolanic Reactivity," *Australian Journal of Applied Sciences,* Vol. 6, No. 3, pp. 316-326, 1955.

147. Ravina, D., "Optimized Determination of PFA (Fly Ash) Fineness with Reference to Pozzolanic Activity," *Cement and Concrete Research,* Vol. 10, No. 4, pp. 573-580, 1980.

148. Weshe, K., and vom Berg, W., "Properties of Fly Ash Used in Germany," pp. 45-52 in 49B.

149. Markowski, G. R.; Ensor, D. S.; Hooper, R. G.; and Carr, R. C., "A Submicron Aerosol Mode in Flue Gas from a Pulverized Coal Utility Boiler," *Environmental Science and Technology,* Vol. 14, No. 11, pp. 1400-1402, 1980.

150. Quann, R. J.; Neville, M.; Janghorbani, M.; Mims, C. A.; and Sarofim, A. F., "Mineral Matter and Trace-element Vaporization in a Laboratory-Pulverized Coal Combustion System," *Environmental Science and Technology,* Vol. 16, No. 11, pp. 776-781, 1982.

151. Montgomery, D. G., "Fly Ash in Cement Paste—Some Effects of Particle Size Distribution," pp. 194-203 in 49B.

152. Clendenning, T. G., and Durie, N. D., "Properties and Use of Fly Ash from a Steam Plant Operating Under Variable Load," *Proceedings, ASTM,* Vol. 62, pp. 1019-1037, 1962.

153. Dhir, R. K.; Apte, A. G.; and Munday, J. G. L., "Effect of In-Source Variability of Pulverized Fuel Ash upon the Strength of OPC/PFA Concrete," *Magazine of Concrete Research,* Vol. 33, No. 117, pp. 199-207, 1981.

154. Ashby, J. B., Discussion of Ref. 29, *Proceedings, American Concrete Institute,* Vol. 78, No. 1, pp. 86-87, 1981.

155. Joshi, R. C., and Ward, M. A., "Cementitious Fly Ashes, Structural and Hydration Mechanism," Vol. III, pp. IV-78-83 in 17B.

156. Yamazaki, K., "Fundamental Studies of Effects of Mineral Fines on Workability of Concrete," *Transactions of the Japan Society of Civil Engineers,* Vol. 84, pp. 98-118; Vol. 85, pp. 15-44, 1962 (in Japanese); (see Ref. 3, in English).

157. Helmuth, R. A., "Structure and Rheology of Fresh Cement Paste," Vol. III, pp. VI-0/1-30 in 17B.

158. Bombled, J. P., "Rheology of Mortars and Fresh Concrete: Studies of the Interstitial Cement Paste," *Ciments, Bétons, Platres, Chaux (Revue des Materiaux de Construction),* No. 688, pp. 137-155, 1974 (in French).

159. Chappuis, J., "Relation of the Rheological Properties of Dense Suspensions to Interparticle Forces," pp. 38-45 in 20B.

160. Powers, T. C., p. 194 in 5B.

161. Vom Berg, W., "Influence of Specific Surface and Concentration of Solids upon the Flow Behavior of Cement Pastes," *Magazine of Concrete Research,* Vol. 31, No. 109, pp. 211-216, 1979.

162. Legrand, C., "A Contribution to the Study of Fresh Concrete Rheology," Materials and Structures, Vol. 5, No. 29, pp. 275-295; No. 30, pp. 379-393, 1972 (in French).

163. Ivanov, Y., and Zacharieva, S., "Influence of Fly Ash on the Rheology of Cement Pastes," Vol. III, pp. VI-103-107 in 17B.

164. Helmuth, R., Discussion of Ref. 142, *Magazine of Concrete Research,* Vol. 33, No. 117, pp. 227-230, 1981.

165. Helmuth, R., "Some Questions Concerning ASTM Standards and Methods of Testing Fly Ash for Use with Portland Cement," *Cement, Concrete, and Aggregates,* ASTM, Vol. 5, No. 1, pp. 103-110, 1983.

166. Stuart, K. D.; Anderson, D. A.; and Cady, P. D., "Compressive Strength Studies on Portland Cement Mortars Containing Fly Ash and Superplasticizers," *Cement and Concrete Research,* Vol. 10, No. 6, pp. 823-832, 1980.

167. Yuan, R. Z.; Jin, S. X.; and Qian, J. C., "Effects of Fly Ash on the Rheology of Fresh Cement," pp. 182-191 in 20B.

168. Helmuth, R., "Water-Reducing Properties of Fly Ashes in Cement Pastes, Mortars, and Concretes: Causes and Test Methods" pp. 723-740 in 58B.

169. Meininger, R. C., "Use of Fly Ash in Air-Entrained Concrete—Report of Recent NSGA-NRMCA Research Laboratory Studies," National Sand and Gravel Association, 32 pp., 1981.

170. Dodson, V. H., "Characterization of Fly Ashes with Respect to Air-Entraining Agent Requirements," presented at Transportation Research Board, 59th Annual Meeting, Session 134 (taped), January 1980.

171a. Committee on Basic Research Pertaining to Portland Cement and Concrete, *Guide to Compounds of Interest in Cement and Concrete Research,* Special Report 127, Highway Research Board, National Research Council-NAS-NAE, 53 pp., Washington, D.C., 1972.

171b. Seligmann, P., and Greening, N. R., "Phase Equilibria of Cement-Water," Vol. II, pp. 179-199, in 6B.

172. Skalny, J. P., and Young, J. F., "Mechanisms of Portland Cement Hydration," Vol. I, pp. II-1/3-44 in 17B.

173. Brunauer, S., and Copeland, L. E., "The Chemistry of Concrete," *Scientific American,* Vol. 210, No. 4, pp. 80-93, 1964.

174. Kantro, D. L.; Brunauer, S.; and Weise, C. H., "Development of Surface in the Hydration of Calcium Silicates," in *Solid Surfaces and the Gas-Solid Interface,* Advances in Chemistry Series 33, pp. 199-219, 1961; Portland Cement Association Research Bulletin 140.

175. Copeland, L. E.; Bodor, E.; Chang, T. N.; and Weise, C. H., "Reactions of Tobermorite Gel with Aluminates, Ferrites, and Sulfates," *Journal of the Portland Cement Association R&D Laboratories,* Vol. 9, No. 1, pp. 61-74, 1967; Portland Cement Association Research Bulletin 211.

176. Copeland, L. E.; Kantro, D. L.; and Verbeck, G., "Chemistry of Hydration of Portland Cement," Vol. 1, pp. 429-465 in 1B; Portland Cement Association Research Bulletin 153.

177. Jawed, I., and Skalny, J. P., "Alkalies in Cement: A Review, II. Effects of Alkalies on Hydration and Performance of Cement," *Cement and Concrete Research,* Vol. 8, No. 1, pp. 37-52, 1978.

178a. Lerch, W., "The Influence of Gypsum on the Hydration and Properties of Portland Cement Pastes," *Proceedings, ASTM,* Vol. 46, pp. 1252-1292, 1946; Portland Cement Association Research Bulletin 12.

178b. Kanare, H. M., and Gartner, E. M., "Optimum Sulfate in Portland Cement," pp. 213-251 in CRP 1984 23B.

179. Lachowski, E. E.; Mohan, K.; and Taylor, H. F. W., "Analytical Electron Microscopy of Cement Pastes, II. Pastes of Portland Cements and Clinkers," *Journal of the American Ceramic Society,* Vol. 63, No. 7-8, pp. 447-452, 1980.

180. Diamond, S., "Cement Paste Microstructure—An Overview at Several Levels," pp. 2-30 in 11B.

181. Powers, T. C., "Physical Properties of Cement Pastes," Vol. II, pp. 577-608 in 1B.

182a. Winslow, D. N., and Diamond, S., "Specific Surface of Hardened Portland Cement Paste As Determined by Small-Angle X-Ray Scattering," Journal of the American Ceramic Society, Vol. 57 (5), pp. 193-197, May 1974.

182b. Copeland, L. E., and Verbeck, G. J., "Structure and Properties of Hardened Cement Paste," Paper II-5 in 9B.

183a. Helmuth, R., and Turk, D. H., "The Reversible and Irreversible Drying Shrinkage of Hardened Portland Cement and Tricalcium Silicate Pastes," *Journal of the Portland Cement Association R&D Laboratories,* Vol. 9, No. 2, pp. 8-21, 1967; Portland Cement Association Research Bulletin 215.

183b. Parrott, L. J., "Effect of Drying History Upon the Exchange of Pore Water with Methanol and Upon Subsequent Methanol Adsorption Behavior in Hydrated Alite Paste," *Cement and Concrete Research,* Vol. 11, pp. 651-658, 1981.

184. Verbeck, G., and Helmuth, R., "Structures and Properties of Cement Pastes," Vol. III, pp. 1-31 in 6B.

185. Takemoto, K., and Uchikawa, H., "Hydration of Pozzolanic Cement," Vol. I, pp. IV-2/1-29 in 17B.

186. Clifton, J. R.; Brown, P. W.; and Frohnsdorff, G., "Reactivity of Fly Ashes with Cement," in *Cements Research Progress 1977,* Ch. 15, pp. 321-341, Columbus, Ohio, American Ceramic Society, 1978.

187. ISO Recommendation R597: Definitions et Terminologie des Ciments, *La Edizione,* 1967 (in French).

188. Lea, F. M., pp. 179, 187-195 in 7B.

189. Taylor, H. F. W., "The Calcium Silicate Hydrates," Vol. 1, pp. 167-232 in 2B.

190. Greening, N. R., "CaO-SiO_2-H_2O Relations in Calcium Silicate Hydrates," Session 139 (taped), Transportation Research Board Annual Meeting, Washington, D.C., January 1978.

191. Lea, F. M., p. 189 in 7B.

192. Wittmann, F. H., "Properties of Hardened Cement Paste," Vol. I, pp. V1-2/1-14 in 17B.

193. Greenberg, S., and Chang, T. N., "Investigation of Colloidal Hydrated Calcium Silicates, II. Solubility Relationships in the Calcium Oxide-Silica-Water System at 25° C," *Journal of Physical Chemistry,* Vol. 69, No. 1, pp. 182-188, 1965; Portland Cement Association Research Bulletin 180.

194. Dron, R., "Experimental and Theoretical Study of the CaO-Al_2O_3-SiO_2-H_2O System," Supplementary Paper 9-5 in 9B.

195. Kalousek, G. L., "Studies of Portions of the Quaternary System Soda-Lime-Silica-Water at 25° C," *Journal of Research,* National Bureau of Standards, Vol. 32, pp. 285-302, 1944.

196. Dron, R., "Les Pouzzalanes et la Pouzzolanicité," *Revue des Materiaux de Construction,* No. 692, pp. 27-30, 1975 (in French).

197. Turriziani, R., "Aspects of the Chemistry of Pozzolans," Vol. 2, pp. 69-86 in 2B.

198. Turriziani, R., "The Calcium Aluminate Hydrates and Related Compounds," Vol. 1, pp. 233-286 in 2B.

199. Malquori, G., "Portland-Pozzolan Cement," Vol. II, pp. 983-1000 in 1B.

200. Kokubu, M., and Yamada, J., "Fly Ash Cements," principal paper III-3, 51 pp. in 9B.

201. Massazza, F., "Chemistry of Pozzolanic Additions and Mixed Cements," principal paper III-6, 65 pp. in 9B; also, *Il Cemento,* Vol. 73, No. 1, pp. 3-39, 1976.

202. Ogawa, K.; Uchikawa, H.; and Takemoto, K., "The Mechanism of the Hydration in the System C3S-Pozzolana," *Cement and Concrete Research,* Vol. 10, No. 5, pp. 683-696, 1980.

203. Verhasselt, A., "Characterization of the Pozzolanicity of Fly Ash," Vol. III, pp. IV-116-121 in 17B.

204. Joshi, R. C. and Rosauer, E. A., "Pozzolanic Activity in Synthetic Fly Ashes, I. Experimental Production and Characterization," *American Society Ceramic Bulletin,* Vol. 52, No. 5, pp. 456-458; "II. Pozzolanic Behavior," pp. 459-463, 1973.

205. Jambor, J., "Hydrationsproducte der Kalk-Puzzolan-Bindemittel," *Zement-Kalk-Gips,* Vol. 52, No. 5, pp. 177-186, 1963 (in German).

206. Sersale, R., and Orsini, P. G., "Hydrated Phases After Reaction of Lime with 'Pozzolanic' Materials or with Blast Furnace Slags," Part IV, pp. 114-121 in 6B.

207. Minnick, L. J., "Reactions of Hydrated Lime with Pulverized Coal Ash," *Minerals and Processing,* Vol. 9, No. 2, pp. 15-20, No. 3, pp. 12-19, 1968.

208. Kondo, R., and Ohsawa, S., "Reactivities of Various Silicates with Calcium Hydroxide and Water," *Journal of the American Ceramic Society,* Vol. 62, No. 9-10, pp. 447-449, 1979.

209. Brunauer, S., and Greenberg, S., "The Hydration of Tricalcium Silicate and Å-Dicalcium Silicate at Room Temperature," Vol. I, pp. 135-165, in 1B; Portland Cement Association Research Bulletin 152.

210. Jawed, I., and Skalny, J., "Hydration of Tricalcium Silicate in the Presence of Fly Ash," pp. 60-70 in 49B.

211. Mohan, K., and Taylor, F. H. W., "Pastes of Tricalcium Silicate with Fly Ash—Analytical Electron Microscopy, Trimethylsilylation and Other Studies," pp. 54-59 in 49B.

212. Plowman, C., and Cabrera, J. G., "The Influence of Pulverized Fuel Ash on the Hydration Reactions of Calcium Aluminates," pp. 71-81 in 49B.

213. Sellevold, E. J.; Badger, D. H.; Jensen, E. K; and Knudsen, T., "Silica Fume-Cement Pastes: Hydration and Pore Structure," *Proceedings, Nordisk Miniseminar on Silica in Concrete,* Cement and Concrete Research Institute, Norwegian Institute of Technology, Trondheim, Norway, 1981.

214. Diamond, S., and Lopez-Flores, F., "Comparative Studies of the Effects of Lignitic and Bituminous Fly Ashes in Hydrated Cement Systems," pp. 112-123 in 49B.

215. Diamond, S.; Ravina, D.; and Lovell, J., "The Occurrence of Duplex Films on Fly Ash Surfaces," *Cement and Concrete Research,* Vol. 10, No. 2, pp. 297-300, 1980.

216. Grandet, J., and Ollivier, J. P., "Orientation of Hydration Products Near Aggregate Surfaces," Vol. III, pp. VII/63-68 in 17B.

217. Terrier, P., and Moreau, M., "Recherche sur le Mecanism de l'Action Pouzzolanique des Cendres Volantes dans le Ciment," *Revue des Materiaux de Construction*, No. 613, pp. 379-396; No. 614, pp. 440-451, 1966 (in French).

218. Bhatty, M. S. Y., and Greening, N. R., "Some Long-Time Studies of Blended Cements with Emphasis on Alkali-Aggregate Reaction," to be published in the *Proceedings of the Seventh International Conference on Alkali-Aggregate Reaction,* Ottawa, Canada, August 18-22, National Research Council Canada, 1986.

219. Thorne, D. J., and Watt, J. D., "Composition and Pozzolanic Properties of Pulverized Fuel Ashes, II. Pozzolanic Properties of Fly Ashes, as Determined by Crushing Strength Tests on Lime Mortars," *Journal of Applied Chemistry,* Vol. 15, pp. 595-604, 1965. See also Ref. 123, 220.

220. Watt, J. D., and Thorne, D. J., "The Composition and Pozzolanic Properties of Pulverized Fuel Ashes, III. Pozzolanic Properties of Fly Ashes as Determined by Chemical Methods," *Journal of Applied Chemistry,* Vol. 16, pp. 33-39, 1966; see also Ref. 123, 219.

221. Feldman, R. F., "Pore Structure Formation During Hydration of Fly Ash and Slag Cement Blends," pp. 124-133 in 49B.

222. Feldman, R. F., "Significance of Porosity Measurements on Blended Cement Performance," pp. 415-433 in 54B.

223. Grutzeck, M. W.; Roy, D. M.; and Scheetz, B. E., "Hydration Mechanisms of High-Lime Fly Ash in Portland Cement Composites," pp. 92-101 in 49B.

224. Feret, R., "The Nature and Progression of Pozzolanic Action," *Revue des Materiaux de Construction,* No. 281, pp. 41-44; No. 282, pp. 85-92; Vol. 288, p. 293, 1933 (in French).

225. Isard, J. O., "The Formation of Spherical Glass Particles on the Lunar Surface," *Proceedings, Second Lunar Science Conference,* Vol. 3, pp. 2003-2008, 1971.

226. Ings, J., "Effect of Thermal History on Fly Ash Reactivity," presented at the Annual Meeting of the American Ceramic Society, Chicago, Paper 21-T-80, 1980.

227. Ravina, D., "Efficient Utilization of Coarse and Fine Fly Ash in Precast Concrete by Incorporating Thermal Curing," *Proceedings, American Concrete Institute,* Vol. 78, No. 3, pp. 194-200, 1981.

228. Valore, R. C., "Laboratory Evaluation of Fly Ash and Other Pozzolans for Use in Concrete Products," pp. 171-192 in 38B.

229. Fratini, N., "Research on Lime Released by Hydration of Cement Paste, II. Suggestion for a Method of Test for the Chemical Evaluation of Pozzolanic Cements," *Annual di Chimica Applicata,* Vol. 40, pp. 461-469, 1950 (in Italian).

230. ISO Recommendation R863: Essai de Pouzzolanicité pour le Controle des Ciments Pouzzolaniques, *La Edizione,* 1968 (in French).

231. Luxan, M. P. de, and Soria, F., "Study and Critical Review of the Pozzolanicity Test," *Cement and Concrete Research,* Vol. 5, No. 5, pp. 461-480, 1975; discussion by Bensted (Ref. 234).

232. Matthews, J. D., and Gutt, W. H., "Research on Pulverized Fuel Ash As a Material for Cement," International Conference on Ash Technology and Marketing, London, pp. 1-21, 1978.

233. Bonomi, F., and Francardi, M. T., "A Further Contribution to Improve the Pozzolanicity Test," *Il Cemento,* Vol. 72, No. 4, pp. 163-172, 1975.

234. Bensted, J., Discussion of Ref. 231, *Cement and Concrete Research,* Vol. 7, No. 4, pp. 461-462, 1977.

235. Sestini, Q., and Santarelli, L., "Research on Pozzolanas, IV. Analyses of Pozzolanas and Pozzolanic Mortars," *Annal di Chimica Applicata,* Vol. 26, No. 12, pp. 533-537, 1936 (in Italian).

236. Lea, F. M., "Investigation on Pozzolanas, I. Pozzolanas and Lime-Pozzolana Mixes," Technical Paper No. 27, Building Research, Station, Great Britain, 63 pp., 1940.

237. Moran, W. T., and Gilliland, J. L., "Summary of Methods for Determining Pozzolanic Activity," ASTM STP 99, pp. 109-125, Philadelphia, American Society for Testing and Materials, 1950.

238. Raverdy, M.; Brivot, F.; Paillere, A. M.; and Dron, R., "Appreciation of Pozzolanic Reactivity of Minor Components," Vol. III, pp. IV/36-41 in 17B.

239. Raask, E., and Bhaskar, M. C., "Pozzolanic Activity of Pulverized Fuel Ash," *Cement and Concrete Research,* Vol. 5, No. 4, pp. 363-376, 1975.

240. Raask, E., "Utilization of Pozzolanic and Cenospheric Ashes," Vol. III, pp. IV/1-6 in 17B.

241. Majumdar, A. J., and Larner, L. J., "The Measurement of Pozzolanic Activity," *Cement and Concrete Research,* Vol. 7, No. 2, pp. 209-210, 1977.

242. Mehta, P. K., "History and Status of Performance Tests for Evaluation of Soundness of Cements," pp. 35-56 in 14B.

243a. Harris, H., Private Communication, 1981.

243b. Schlorholtz, S., and Demirel, T., "Autoclave Expansion of Portland Cement-Fly Ash Pastes," pp. 85-93 in 55B.

244. Rosa, J., "Sound Cements with High MgO Content," *Zement-Kalk-Gips,* Vol. 9, pp. 460-470, 1965; English Translation: Library Communication No. 1362, Building Research Station, Great Britain.

245. "Utilization of High Magnesia Limestone in the Manufacture of Cement," RR-1, Cement Research Institute of India, 24 pp., 1969.

246. Rehsi, S. S., and Majumdar, A. J., "High-Magnesia Portland Cements: Studies on Cements Prepared with Reagent-Grade Chemicals," *Magazine of Concrete Research,* Vol. 21, No. 67, pp. 67-78, 1969.

247. Speakman, K., and Majumdar, A. J., "Synthesis and Characterization of Magnesia-Bearing Phases Related to High-Magnesia Cement Hydration Products," *Proceedings of the Tenth Conference on the Silicate Industry,* Budapest, 1970.

248. Gaze, M. E., and Smith, M. A., "High Magnesia Cements, Part 2: The Effect of Hydraulic and Non-Hydraulic Admixtures on Expansion," *Cement Technology,* Vol. 5, No. 2, pp. 291-295, 1974.

249. Rehsi, S. S., and Garg, S. K., "Room Temperature Stabilization of High Magnesia Cement with Fly Ash," *Zement-Kalk-Gips,* Vol. 28, No. 2, pp. 84-87, 1975 (in German).

250. Rosa, J., "The Mechanism of the Stabilization of High-Magnesia Portland Cement," *Proceedings of the Eighth Conference on the Silicate Industry,* Budapest, pp. 1965.

251. Kurdowski, W., "Expansive Cements," Vol. I, pp. V-2/1-11 in 17B.

252. Carles-Giberures, A.; Thenoz, B.; and Vaguier, A., "Hydration Mechanism of a Fly Ash Produced from Calcareous Lignite," Vol. III, pp. IV/53-59 in 17B.

253. Richards, C. W., and Helmuth, R. A., "Expansive Cement Concrete—Micromechanical Models for Free and Restrained Expansion," CE-TR-191, Dept. of Civil Engineering, Stanford University, 36 pp., 1975 (NTIS PB 242302).

254. Lea, F. M., p. 346 in 7B.

255. Mather, K., "Factors Affecting Sulfate Resistance of Mortars," Miscellaneous paper SL-80-17, 18 pp., Structures Laboratory, U.S. Army Engineers, Waterways Experiment Station, Vicksburg, Miss, 1980. See also Vol. IV, pp. 580-585 in 17B and pp. 63-77 in *Sulfate Resistance of Concrete,* ACI SP-77, Detroit, American Concrete Institute, 1982.

256a. Kalousek, G. L., and Benton, E. J., "Mechanisms of Sea Water Attack on Cement Pastes," *Proceedings, American Concrete Institute,* Vol. 67, No. 2, pp. 187-192, 1970.

256b. Dunstan, E. R., Jr., "A Spec Odyssey—Sulfate Resistant Concrete for the 1980's," *Sulfate Resistance of Concrete,* ACI SP-77, Detroit, American Concrete Institute, 1982.

257a. Wolochow, D., "Determination of the Sulfate Resistance of Portland Cement," *Proceedings, ASTM,* Vol. 52, pp. 252-263, 1952.

257b. Rosner, J. C.; Chehovits, J. G.; and Warburton, R. G., "Sulfate Resistance of Mortars using Fly Ash As a Partial Replacement for Portland Cement," pp. 37-56 in 51B.

258. Revay, M., and Kovacs, R., "Some Aspects and a New Method of Evaluating the Sulfate Resistance of Cements," Vol. III, pp. VII/34-39 in 17B.

259. Mehta, P. K., and Gjorv, O. E., "A New Test Method for Sulfate Resistance of Cements," *Journal of Testing and Evaluation,* Vol. 2, No. 6, pp. 510-515, 1974.

260. Brown, P. W., "An Evaluation of the Sulfate Resistance of Cements in a Controlled Environment," *Cement and Concrete Research,* Vol. 11, No. 5/6, pp. 719-727, 1981.

261. Stark, D., "Alkali-Silica Reactivity: Some Reconsiderations," *Cement, Concrete, and Aggregates,* ASTM, Vol. 2, No. 1, pp. 92-94, 1980.

262. Hobbs, D. W., "Influence of Pulverized-Fuel Ash and Granulated Blast Furnace Slag Upon Expansion Caused by the Alkali-Silica Reaction," *Magazine of Concrete Research,* Vol. 34, No. 119, pp. 83-94, 1982.

263. Gutt, W.; Nixon, P. J; and Gaze, M. E., "Fly Ash and Alkali Aggregate Reaction," Vol. III, pp. VII/110-114 in 17B.

264. Powers, T. C., and Steinour, H., "An Interpretation of Published Researches on the Alkali-Aggregate Reaction, I. The Chemical Reactions and Mechanism of Expansion, II. A Hypothesis Concerning Safe and Unsafe Reactions with Reactive Silica in Concrete," *Proceedings, American Concrete Institute,* Vol. 51, No. 6, pp. 497-516; No. 8, pp. 785-812, 1955; Portland Cement Association Research Bulletin 55.

265. Struble, L., "Swell and Other Properties of Synthetic Alkali Silica Gels," M.S. Thesis, Purdue University, 1979.

266. Bhatty, M. S. Y., and Greening, N. R., "Interaction of Alkalies with Hydration and Hydrated Calcium Silicates," pp. 87-112 in 12B.

267. Rayment, P. L., "The Effect of Pulverized-Fuel Ash on the C/S Molar Ratio and Alkali Content of Calcium Silicate Hydrates in Cement," Cement and Concrete Research, Vol. 12, No. 2, pp. 133-140, 1982.

268. Pepper, L., and Mather, B., "Effectiveness of Mineral Admixtures in Preventing Excessive Expansion of Concrete Due to Alkali-Aggregate Reaction," *Proceedings, ASTM,* Vol. 59, pp. 1178-1202; discussions, p. 1202, 1959.

269. Pepper, L., "Influence of Alkali Content of Fly Ash on Effectiveness in Preventing Expansion of Concrete," Technical Report No. 6-627, U.S. Army Engineers, Waterways Experiment Station, Vicksburg, Miss., 1963.

270. Stark, D., "Alkali-Silica Reactivity in the Rocky Mountain Region," pp. 235-243 in 12B.

271. Dunstan, E. R., Jr., "The Effect of Fly Ash on Concrete Alkali-Aggregate Reaction," *Cement, Concrete, and Aggregates,* CCAGDP, Vol. 3, No. 2, pp. 101-104, 1981.

272. Hobbs, D. W., "Influence of Mix Proportions and Cement Alkali Content upon Expansion Due to the Alkali-Silica Reaction," Technical Report 534, 31 pp., Cement and Concrete Association, Great Britain, 1980.

273. Hobbs, D. W., "The Alkali-Silica Reaction—A Model for Predicting Expansion in Mortar," *Magazine of Concrete Research,* Vol. 33, No. 117, pp. 208-220, 1981.

274. Gaze, M. E., and Nixon, P. J., "The Effect of PFA Upon Alkali-Aggregate Reaction," *Magazine of Concrete Research,* Vol. 35, No. 123, pp. 107-110, 1983.

275. Mather, B., "New Concern Over Alkali-Aggregate Reaction," NRMCA Publication No. 149, 20 pp., National Ready Mixed Concrete Association, Silver Spring, Md. 1975.

276. Bakker, R. F. M., "Permeability of Blended Cement Concretes," pp. 589-605 in 54B.

277. Short, N. R., and Page, C. L., "The Diffusion of Chloride Ions Through Portland and Blended Cement Pastes," Silicates Industriels, Vol. 47, No. 10, pp. 237-240, 1982.

278. Kasai, Y.; Matsui, I.; Fukushima, Y.; and Kamohara, A., "Air Permeability and Carbonation of Blended Cement Mortars," pp. 435-451 in 54B.

279. Gautefall, O., and Vennesland, O., "Effect of Cracks on the Corrosion of Embedded Steel in Silica-Concrete Compared to Ordinary Concrete," pp. 17-28 in *Nordic Concrete Research,* Publication No. 2, The Nordic Concrete Federation, Oslo, 1983.

280. Mehta, P. K., "A Simple Test for Evaluation of Quality of Fly Ashes and Other Pozzolans for Use in the Cement and Concrete Industry," pp. 268-277 in 51B.

281. Fagerlund, G., "Predicting the Service Life of Concrete Structures," pp. 187-198 in *Characterization and Performance Prediction of Cement and Concrete,* J. F. Young, Ed., Engineering Foundation Conference Proceedings, United Engineering Trustees, New York, 1983.

Bibliography

Books, Proceedings, and Major Reports

Cement and Concrete

1B. *Proceedings, Fourth International Symposium on the Chemistry of Cement,* Washington D.C., 1960, National Bureau of Standards, Monograph 43, Vols. I and II, 1125 pp., Washington, D.C., U.S. Dept. of Commerce, 1962.

2B. Taylor, H. F. W., Ed., *The Chemistry of Cements,* Vol. 1, 460 pp., Vol. 2, 409 pp., New York, Academic Press, 1964.

3B. Woods, H., *Durability of Concrete Construction,* ACI Monograph No. 4, 187 pp., Detroit, American Concrete Institute and University of Iowa, Ames, Iowa, 1968.

4B. Swenson, E. G., Ed., *Performance of Concrete,* 243 pp., Toronto, University of Toronto Press, 1968.

5B. Powers, T. C., *The Properties of Fresh Concrete,* 664 pp., New York, John Wiley & Sons, 1968.

6B. *Proceedings, Fifth International Symposium on the Chemistry of Cement,* Tokyo, 1968, Part I, 384 pp.; Part II, 513 pp.; Part III, 601 pp.; Part IV, 519 pp., Tokyo, Cement Association of Japan, 1969.

7B. Lea, F. M., *The Chemistry of Cement and Concrete,* 3rd ed., 727 pp., New York, Chemical Publishing Co., 1971.

8B. Neville, A. M., *Properties of Concrete,* 2nd ed., 687 pp., London, Pitman Publishing, 1975.

9B. *Sixth International Congress on the Chemistry of Cement,* Moscow, 1974, English preprints, 27 principal papers, 190 supplementary papers, Moscow, 1974.

10B. Portland Cement Association, *Energy Conservation Potential in the Cement Industry,* Conservation Paper Number 26, 309 pp., Washington, D.C., Federal Energy Administration, 1975 (NTIS PB 245 159/AS).

11B. *Hydraulic Cement Pastes: Their Structure and Properties,* Proceedings of a Conference held at University of Sheffield, April 8-9, 1976, 334 pp., Wexham Springs, Cement and Concrete Association, 1976.

12B. *Proceedings, Fourth International Conference on the Effects of Alkalies in Cement and Concrete,* June 4-7, 1978, Purdue University, Publication No. CE-MAT-1-78, 376 pp., West Lafayette, Ind., Purdue University Press, 1978.

13B. Rixom, M. R., *Chemical Admixtures for Concrete,* 234 pp., London, E&FN Spon Ltd., 1978.

14B. Mehta, P. K., Ed., *Cement Standards—Evolution and Trends,* ASTM STP 663, 107 pp., Philadelphia, American Society for Testing and Materials, 1978.

15B. *Significance of Tests and Properties of Concrete-Making Materials,* ASTM STP 169B, 872 pp., Philadelphia, American Society for Testing and Materials, 1978.

16B. Malhotra, V. M.; Berry, E. E.; and Wheat, T. A., *Superplasticizers in Concrete,* Proceedings of an International Symposium, Ottawa, Canada, May 1978, 2 Vols., 801 pp., Detroit, American Concrete Institute, 1979.

17B. *Proceedings, Seventh International Congress on the Chemistry of Cement,* Paris 1980, 4 Vols., Paris Editions Septima, 1980.

18B. *The Status of Cement and Concrete R&D in the United States,* Publication NMAB-361, 103 pp., Washington, D.C., National Materials Advisory Board, National Research Council, 1980 (NTIS PB 80175672).

19B. Mindess, S., and Young, J. F., *Concrete,* 671 pp., New York, Prentice Hall, 1981.

20B. Skalny, J. P., Ed., *Effect of Surface and Colloid Phenomena on Properties of Fresh Concrete,* Proceedings, Symposium M, Materials Research Society, Annual Meeting, November 1982, 233 pp., University Park, Pa, Materials Research Society, 1982.

21B. *Annual Book of ASTM Standards Vol. 04.01* [*Part 13 before 1983*]/ *Cement; Lime; Gypsum;* 723 pp., Philadelphia, American Society for Testing and Materials, 1986.

22B. *Annual Book of ASTM Standards Vol. 04.02* [*Part 14 before 1983*]/ *Concrete and Mineral Aggregates;* 945 pp., Philadelphia, American Society for Testing and Materials, 1986.

23B. Young, J. F., Ed., *Cements Research Progress,* Annual survey of the literature on the science of cements, Columbus, Ohio, American Ceramic Society, 1974-1985.

24B. *ACI Manual of Concrete Practice 1983 Part 1—Materials and General Properties of Concrete,* Detroit, American Concrete Institute, 1983.

Coal and Coal Combustion

25B. Pettijohn, F. J., *Sedimentary Rocks,* 549 pp., New York, Harper and Bros., 1957.

26B. Gary, M; McAfee, R., Jr.; and Wolf C. L., Eds., *Glossary of Geology,* Washington, D.C., American Geological Institute, 1972.

27B. Moorhouse, W. W., *The Study of Rocks in Thin Section,* 514 pp., New York, Harper and Bros., 1959.

28B. Field, M. A.; Gill, D. W.; Morgan, B. B.; and Hawksley, P. G. W., *Combustion of Pulverized Coal,* 477 pp., Leatherhead, British Coal Utilization Research Association, 1967.

29B. Yaverbaum, L., *Fluidized Bed Combustion of Coal and Waste Materials,* 267 pp., Park Ridge, N.J., Noyes Data Corporation, 1977.

30B. Smoot, L. D., and Pratt, D. T., Eds., *Pulverized-Coal Combustion and Gasification,* 333 pp., New York, Plenum Press, 1979.

Fly Ash

31B. Jarrige, A., *Les Cendres Volantes,* 216 pp., Paris, Editions Eyrolles, 1971.

32B. Dismukes, E. B., *Conditioning of Fly Ash with Sulfur Trioxide and Ammonia,* Final Report (1970-1975), Southern Research Institute to EPA Office of R&D, SORI-EAS-75-311, 169 pp., Chattanooga, Tennessee Valley Authority, 1975.

33B. Ray, S. S., and Parker, F. G., *Characterization of Ash from Coal-Fired Power Plants,* EPA 600/7-77-010, 123 pp., Chattanooga, Tennessee Valley Authority, 1977.

34B. Powell, R. A., and Spicer, W. E., *Characterization of Fly Ash and Related Metal Oxides Using Auger Electron Spectroscopy,* EPRI-FP-708, 124 pp., Palo Alto, Calif., Electric Power Research Institute, 1978.

35B. GAI Consultants, *Coal Ash Disposal Manual,* FP-1257, 250 pp., Palo Alto, Calif., Electric Power Research Institute, 1979.

Fly Ash Utilization in Concrete

36B. Abdun-Nur, E. A., *Fly Ash in Concrete—An Evaluation,* Bulletin 284, 135 pp., Washington, D.C., Highway Research Board, 1961.

37B. Faber, J. H.; Capp, J. P.; and Spencer, J. D., Compilers, *Proceedings, Symposium on Fly Ash Utilization,* Information Circular IC 8348, 345 pp., Washington, D.C., U.S. Dept. of the Interior, Bureau of Mines, 1967.

38B. Faber, J. H.; Capp, J. P.; and Spencer, J. D., Compilers, *Proceedings, Second Ash Utilization Symposium,* Information Circular IC 8488, 345 pp., Washington, D.C., U.S. Dept. of the Interior, Bureau of Mines, 1970.

39B. Faber, J. H.; Eckard, W. E.; and Spencer, J. D., Compilers, *Proceedings, Third International Ash Utilization Symposium,* Information Circular IC 8640, 317 pp., Washington, D.C., U.S. Dept. of the Interior, Bureau of Mines, 1974.

40B. Faber, J. H.; Babcock, A. W.; and Spencer, J. D., Compilers, *Proceedings, Fourth International Ash Utilization Symposium,* MERC/ Sp-76-4 (CONF-760322), 688 pp., U.S. Energy Research and Development Administration, Morgantown, W.Va., 1976.

41B. Berry, E. E., *Fly Ash for Use in Concrete, Part I—A Critical Review of the Chemical, Physical, and Pozzolanic Properties of Fly Ash,* CANMET Report 76-25, 68 pp., Ottawa, Canada Centre for Mineral and Energy Technology, August 1976.

42B. Berry, E. E., and Malhotra, V. M., *Fly Ash for Use in Concrete, Part II—A Critical Review of the Effects of Fly Ash on the Properties of Concrete,* CANMET Report 78-16, 51 pp., Ottawa, Canada Centre for Mineral and Energy Technology, 1978.

43B. Torrey, S., *Coal Ash Utilization, Fly Ash, Bottom Ash, and Slag,* 370 pp., Park Ridge, N.J., Noyes Data Corporation, 1978.

44B. Price, J. D.; Troop, P.; and Gershman, H. W., *Potential for Energy Conservation Through the Use of Slag and Fly Ash in Concrete,* DOE EC-77-C-03-1699, SAN-1699-T1, 368 pp., Washington, D.C., U.S. Dept. of Energy, 1978.

45B. *Properties and Use of Fly Ash in Portland Cement Concrete,* Technical Report CR-79-2, Knoxville, Singleton Materials Engineering Laboratory/Tennessee Valley Authority, February 1979.

46B. Faber, J. H., and Babcock, A. W., Compilers, and Spencer, J. D., and Whieldon, C. E., Jr., Eds., *Proceedings, Fifth International Ash Utilization Symposium,* METC/SP-79/10, Parts 1 and 2, 950 pp., U.S. Dept. of Energy, Morgantown, W.Va., 1979.

47B. *Symposium on the Utilization of Pulverized Fuel Ash,* Council for Scientific and Industrial Research, Pretoria, South Africa, 1979.

48B. Rivera, R., Ed., *Proceedings, Fifth International Symposium on Concrete Technology: Pozzolans and Marine Environment,* Monterrey, Mexico, March 1981, 29 papers, Universidad Autonoma de Nuevo Leon, 1981.

49B. Diamond, S., Ed., *Effects of Fly Ash Incorporation in Cement and Concrete,* Proceedings, Symposium N, Materials Research Society Annual Meeting, November 1981, 331 pp., University Park, Pa., Materials Research Society, 1981.

50B. Bakker, W. T., Ed., *Workshop Proceedings: Research and Development Needs for Use of Fly Ash in Cement and Concrete,* March 1982, CS-2616-SR, Palo Alto, Calif., Electric Power Research Institute, 1982.

51B. Halow, J. S., and Covey, J. N., *The Challenge of Change—Sixth International Ash Utilization Symposium,* DOE/METC/82-52 Vol. 1, 491 pp; Vol. 2, 426 pp., U.S. Dept. of Energy, Morgantown, W.Va., and National Ash Association, 1982.

52B. Casens, A. R., Ed., *International Symposium on the Use of Pulverized Fuel Ash in Concrete,* April 1982, 343 pp., Leeds, University of Leeds, 1982.

53B. Dunstan, E. R., Jr., Ed., *Denver Fly Ash Symposium,* January 1983, 21 papers, Lakewood, Colo., Dunstan, Inc., 1983.

54B. Malhotra, V. M., Ed.,*Fly Ash, Silica Fume, Slag and Other Mineral By-Products in Concrete,* SP-79, 2 Vols., 1182 pp., Detroit, American Concrete Institute, 1983.

55B. McCarthy, G. J., and Lauf, R. J., Eds., *Fly Ash and Coal Conversion By-Products: Characterization, Utilization, and Disposal I,* Materials Research Society Symposia Proceedings, Vol. 43, 272 pp., Pittsburg, Materials Research Society, 1985.

56B. McCarthy, G. J.; Glasser, F. P.; and Roy, D. M., Eds. *Fly Ash and Coal Conversion By-Products: Characterization, Utilization, and Disposal II,* Materials Research Society Symposia Proceedings, Vol. 65, 325 pp., Pittsburg, Materials Research Society, 1986.

57B. Berry, E. E. and Malhotra, V. M., *Fly Ash in Concrete* • Les Cendres Volantes dans le Béton, CANMET SP 85-3, Ottawa, Canadian Government Publishing Centre, 1986.

58B. Malhotra, V. M., Ed., *Fly Ash, Silica Fume, Slag, and Natural Pozzolans in Concrete,* SP-91, 2 Vols., 1609 pp., Detroit, American Concrete Institute, 1986.

Index